Safe at Home with Assistive Technology

Ingrid Kollak
Editor

Safe at Home with Assistive Technology

 Springer

Editor
Ingrid Kollak
Alice Salomon University of Applied
  Science Berlin
Berlin
Germany

ISBN 978-3-319-82687-5   ISBN 978-3-319-42890-1   (eBook)
DOI 10.1007/978-3-319-42890-1

Printed on acid-free paper

This Springer imprint is published by Springer Nature
The registered company is Springer International Publishing AG
The registered company address is: Gewerbestrasse 11, 6330 Cham, Switzerland

Acknowledgements

As initiator, coordinator, reviewer and (co)author, I cooperated with all people working directly or supportive of the book. With all of them I talked in meetings or via mails and enjoyed this privilege. I want to thank the authors for the diverse and well-accomplished chapters reflecting their enormous work and experience. I am grateful to Steven Kranz for his support revising the citations and biographical data and improving the graphs as well as for generating the list of literature and uploading all texts. Special thanks go to Fiona Rintoul (http://weepress.co.uk) who did some of the translations and many reviews, helping to make this an enjoyable book to read.

End of August 2016   Ingrid Kollak

Contents

Chapter 1
Prerequisites: Assistive Technologies Between User Centered Assistance and 'Technicalization'

Ingrid Kollak

Interest in health has grown enormously in recent years. This development is due to a change in perspective. For while the emphasis in the previous century was above all on medical protection in the case of an illness or accident, today it is on maintaining health into old age. The causes of this changed understanding of health lie in demographic and epidemiological changes; people are living longer and becoming ill in a different way, which is to say that what were acute illnesses are today chronic conditions.

Promotion of health, which today takes centre, covers the public as well as the personal health care sectors. Trading under the label of promoting health are offerings that range from publicly financed prevention, rehabilitation and care measures, through corporate health management schemes to privately financed beauty, fitness and wellness treatments.

Terms such as empowerment, customer focus and participation highlight this altered understanding of health and health care. They are supposed to spell the end of the old, patriarchal way of directing and allocating medical and care provision, in which patients endured their prescribed treatment patiently and uncritically. Nonetheless, this normative description of active, responsible, self-directed participants in a health market makes it clear that contemplation is not rated very highly, there is no obligation to treat customers equally, an assumption has been made that this newly won power will be successfully used and no one will decline to participate (Bröckling et al. 2004).

This altered understanding of health has created a continuously growing healthcare market. In 2011, 12 % of all gainfully employed people worked in the health care sector—12 times as many as in the chemicals industry. The outpatient

I. Kollak (✉)
Alice Salomon University of Applied Sciences Berlin (D), Berlin, Germany
e-mail: kollak@ash-berlin.eu

© The Author(s) 2017
I. Kollak (ed.), *Safe at Home with Assistive Technology*,
DOI 10.1007/978-3-319-42890-1_1

sector, where nearly 44 % of healthcare staff work, predominates (Federal Statistical Office Wiesbaden 2016). Health expenditure easily doubled in the period 1992–2014. According to The Information System of the Federal Health Monitoring (2016), a total of Euro 159 billion was spent on health in Germany in 1992 and a total of Euro 328 billion in 2014. That equated to 9.4 % of gross domestic product (GDP) in 1992 and 11.2 % in 2014. Expenditure per head of Euro 1,972 in 1992 had risen to Euro 4,050 by 2014.

A characteristic of an expanding, profit-oriented healthcare market is that it views everyday and age-related problems and mental health problems as objects of medical diagnosis and therapy in order to treat everyone and everything with medicine, operations and cures.

These two tendencies to economise and medicalise the health care markets also influence the development of assistive technology products. However, the potential for economic gain is widely divergent. As part of the Lower Saxony Research Network Design of Environments for Ageing (GAL), Erdmann and Schweigert (2012) asked about 2,000 people in Lower Saxony about their willingness to pay for AAL products. Their research results a low willingness to pay of less than Euro 20 a month. Fachinger et al. (2012) took a different perspective on the market for AAL products under the auspices of the same research network (GAL). They calculated a sales potential for AAL products correlated to the age of the main funder of a household that had a defined price and comfort level as well as offering products from technical assistance systems. These assistance systems came from the areas of: housekeeping and provisions, security and privacy, communication and social circle, as well as health and care. They assume a plausible spectrum of Euro 10.1 billion if the main funder is 75 years old or older to 87.2 billion if the main funder is 50 or older. While Erdmann and Schweigert (2012) established a statistically significant connection between a household's net income and willingness to pay, Fachinger et al. (2012) assumed a connection between technical affinity and age (experience with technology throughout the whole of life and access to technology with increasing age).

In view of the economisation and medicalisation situation, the question arises as to which direction further development in the assistive technology industry will now take.

Alongside the flood of all kinds of technical aids catalogued on the Internet and offered in newspaper supplements, there is a category of assistive technology products, which, because of high development costs and a sometimes limited user group, is looking at this development process very carefully. For not every product is for the mass market in the way that mobile phones or the Internet are.

In this sector, it can be seen that development is oriented ever more strongly towards the user. Usability engineering (Nielsen 1994) marked the start of planning, development and manufacturing of technical products geared towards an imaginary user—which actual users first clapped eyes on when they were on the market. In this way, the 'ux' or user experience (Kuniavsky 2003) wasn't taken account of late and couldn't influence the product until it was redesigned. Greater orientation towards actual users ensued in a bid to shorten the costly and time-consuming path to an accepted product: user-oriented design (Lai et al. 2006). Users were supposed to test prototype versions in practical situations before a product was finalised. A more systematic user orientation happens when users are involved in the whole planning, development and design process: user-centred design (Garrett 2010). In this approach, a product is tracked from engineering, design and application in practice through to maintenance together with users. This way of working requires two things: collaboration between users and their social circle interdisciplinary teamwork. It is known from existing research and development work that many assistive systems need to be accepted not just by their users but also by the people around them. For informal or professional help is often needed to use and maintain the systems. A good example is gaze control. Equally, it has been shown that multidisciplinary teams are needed to plan and complete an assistive technology product.

This book supports as comprehensive as possible an involvement by users in all phases of product development. It gathers together chapters from teams that reflect all steps from needs assessment to planning, through development and design, as well as daily use and maintenance, to evaluation. These teams bring together various perspectives from the fields of architecture, design, (electrical) engineering, ergonomics, gerontology, health science, informatics, management, nursing, physiotherapy, rehabilitation and social science.

Our book is aimed at people who use assistive technology products and their social circle. It is also written for health professionals who introduce and prescribe assistive technology. The book also addresses designers and engineers who develop assistive technology and the distributors who sell these products. We want to give them all an understanding of the range of successful applications, as well as conveying a structured procedure whereby clients and products can be successfully matched.

The book brings together many strands of research and details how elderly, handicapped and people with short-term illnesses can manage their daily life at home with the help of assistive technology. It addresses common problems, options and limitations, as well as common requirements of the above-mentioned group— including their informal and professional carers. The chapters focus on safety at home through technological solutions, information and telemedicine as well as communication and entertainment. The authors discuss solutions from an ethical, technical and social perspective.

Readers from various disciplines with questions about the different phases of product development can find useful information in this book. The chapters range from presentations of successful and outstanding products through evaluation of different product groups to scientific reflection. All the chapters are focused on users and provide practical tips. They use photographs to illustrate how users and technology can work successfully together to ensure safety. The texts present graphs and tables from recent studies that provide facts and evidence about problems, requirements and solutions. The book's reference section provides the authors' CVs, lists all references used in the chapters and a comprehensive list of literature and websites for further reading.

References

Bröckling U, Krasmann S, Lemke T (eds) (2004) Glossar der Gegenwart. Suhrkamp, Frankfurt am Main

Erdmann B, Schweigert H (2012) Zahlungsbereitschaft für AAL-produkte. tagungungsband technik für ein selbstbestimmtes leben—5. Deutscher AAL-Kongress 24.01.2012—25.01.2012 in Berlin

Fachinger U, Koch H, Henke KD et al. (2012) Ökonomische potentiale altersgerechter assistenzsysteme. Ergebnisbericht (BMBF)

Federal Statistical Office Wiesbaden (Statistisches Bundesamt Wiesbaden) (2016): Gesundheit. Personal 2014, Fachserie 12 Reihe 7.3.1. Available via DIALOG. https://www.destatis.de/DE/Publikationen/Thematisch/Gesundheit/Gesundheitspersonal/PersonalPDF_2120731.pdf?__blob=publicationFile. Accessed 17 Aug 2016

Garrett JJ (2010) Elements of user experience: the user-centered design for the web and beyond, 2nd edn. New Riders, Berkeley

Kuniavsky M (2003) Observing the user experience: a practitioner's guide to user research. Morgan Kaufmann, Burlington

Lai HH, Lin YC, Yeh CH et al (2006) User-oriented design for the optimal combination on product design. Int J Prod Econ 100(2):253–267. doi:10.1016/j.ijpe.2004.11.005

Nielsen J (1994) Usability engineering. Morgan Kaufmann, Burlington

The Information System of the Federal Health Monitoring (Gesund-heits-bericht-erstat-tung des Bundes) (2016) Gesundheitsausgaben in Deutschland als Anteil am BIP und in Mio. € (absolut und je Einwohner). Gliederungsmerkmale: Jahre. http://www.gbe-bund.de/oowa921-install/servlet/oowa/aw92/dboowasys921.xwdevkit/xwd_init?gbe.isgbetol/xs_start_neu/&p_aid=i&p_aid=64359188&nummer=522&p_sprache=D&p_indsp=-&p_aid=64359717, Accessed 17 Aug 2016

Chapter 2
Living Safely and Actively in and Around the Home: Four Applied Examples from Avatars and Ambient Cubes to Active Walkers

Martin Biallas, Edith Birrer, Daniel Bolliger, Andreas Rumsch, Rolf Kistler, Alexander Klapproth and Aliaksei Andrushevich

Abstract Assistive technology may support people in staying longer independent and improving the quality of life inside and outside of their homes. However, bringing all necessary aspects together to finally bring a product successfully to the market is challenging, as many researchers in the field have experienced over the last years. It means to tackle the technical issues arising, very closely study the potential customers getting to know their needs, their barriers of acceptance and knowing how to creating value for them and to eventually market, sell and ship the final product to them. We believe that only a joint effort of an interdisciplinary team involving technology, human sciences and business partners will have a chance of success. The following pages document four successful and promising AAL projects and their teams, results and experiences on the way to get there.

Keyword Design thinking · User-centred design · Assistant · Dementia · Informal carers · Active Walker

The original version of this book was revised: Chapter 2 has been updated with the co-author "Aliaksei Andrushevich". The erratum to this chapter is available at 10.1007/978-3-319-42890-1_12

M. Biallas · E. Birrer · D. Bolliger · A. Rumsch · R. Kistler (✉) · A. Klapproth · A. Andrushevich
Engineering and Architecture (CH), iHomeLab,
Lucerne University of Applied Sciences and Arts, Luzern, Switzerland
e-mail: rolf.kistler@hslu.ch

A. Andrushevich
e-mail: aliaksei.andrushevich@hslu.ch

5

2.1   Introduction

Inventing assistive technology (AT) and services that let people stay at home longer and maintain their independence and individual quality of life is a meaningful but challenging task. For various reasons, it was a long, rocky road to get the first products to the people in need. In Europe, AT started about a decade ago with a range of innovative sensor and web technologies. It was a technically driven endeavour mainly initiated by engineers and computer scientists at the dawn of the emerging Internet of Things. Aside from the fact that these technological solutions were not really ready to be implemented as real-life active and assistive living (AAL) products at that time, one of the main issues with these first projects was lack of in-depth knowledge about the real needs of the end users: elderly people themselves and their informal and formal carers.

Having identified the reasons for poor acceptance of early solutions, project teams became more interdisciplinary, involving end-user organisations, psychologists, gerontologists, usability experts, designers and others in a more user-centred, end-user-driven development process. As the technology became more stable and sound, the results of the projects, though still at prototype stage, gained more visibility and acceptance because the people who would use the technology were involved from the beginning and solutions were tested with them in realistic field trials. However, assistive solutions didn't really break through, although they were promoted as having a great future in a huge emerging market. It soon became clear in the community that even a technologically sound solution for which there is demand among the intended target group won't work if the business model and value generated around it do not guarantee sustainable commercial success.

In many respects, this was one of the most challenging tasks for developers of AAL solutions. It meant involving other stakeholders from different sectors such as health care providers, the high-tech industry, private service providers and distributors (for example telcos), public authorities, and national and local governments across Europe, which is very diverse. They all matter and need to put their weight behind rising start-ups to help them bringing these innovations to the market. In particular, incorporating innovative products—or rather classes of products—that provide new but necessary service models into the existing care environments and processes, has proven challenging.

Despite all the challenges AAL product developers face, we still believe—some of us as informal carers ourselves—that there is a need, a market and a way to bring these solutions to people. However, this will require a joint effort by all disciplines covering technology, human sciences and business. This article documents four successful and promising AAL projects that go in this direction, reporting on their teams, results and experiences.

2.2   DALIA—Assistant for Daily Life Activities at Home

iHomeLab—Lucerne University of Applied Sciences and Arts, exthex GmbH, Virtual Assistant bv, TP Vision Belgium, Graz University of Technology, Upper Austria University of Applied Sciences—Institute of Applied Health and Social

Sciences, Volkshilfe Steiermark—gemeinnützige Bertriebs GmbH, terzStiftung, Woonzorg—en dientencentrum 't Dijkhuis, Stëftung Hëllef Doheem

2.2.1 Overview and Aims

At some point in life, an older person may need help to manage the daily activities and his or her personal health at home. In most cases, relatives do the biggest share of the caring. But the new situation causes substantial changes in the lives of the people involved and brings many limitations and personal burdens. Our solution, DALIA, is designed to solve that problem. We have developed a new integrated support system designed to help both older adults as primary end users and their informal carers as secondary users. The solution connects them and also provides them with easy access to formal care and medical services.

There are two main challenges connected with entering the AAL market. The first is user acceptance of AAL applications and their usability. Many older people are either not used to technical devices or physically unable to operate them without help. DALIA uses a virtual personal assistant with a human appearance that supports speech interaction to hide this technical complexity. The assistant is accompanied by simple user interfaces that look the same across different platforms. The second challenge is that AAL solutions tend to focus on a single area (for example, fall detection) and use specialised devices (for example, wristbands). If broader care is needed, this can lead to multiple systems being installed. Aside from the high costs, nobody really wants several different systems at home. DALIA is designed to run on consumer devices found in many contemporary homes already. It seamlessly integrates different service modules in a single application that can run on those devices. In this respect DALIA can provide a cost-effective, holistic and personalised solution for older people and those involved in their care.

DALIA is an essential step towards efficient care support that is economically accessible for everyone.

2.2.2 Implementation

DALIA provides an integrated, individually tailored home system, based on a virtual personal assistant, that supports and empowers older adults to live independently at home and care for themselves for as long as possible. It allows informal carers, formal care providers and medical services to connect with the people they care for and provides them with new and powerful collaboration features.

DALIA relies on current technology and integrates existing consumer devices (smartphone, tablet and TV) to provide a wide range of AAL modules that support older people in their daily life in a single application. The primary users or their informal carers can activate the modules that suit and help them and deactivate modules they won't use. Modules include:

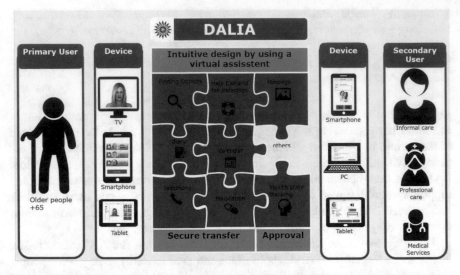

Fig. 2.1 Dalia system overview. *Source* DALIA consortium

- *collaboration* (video calls, sending messages)—enables easy exchange with family and friends, thereby supporting participation in social life
- *calendar*—used as a shared planning aid for appointments (for example, doctor's appointments) and generates automatic reminders
- *diary*—used to support an active lifestyle, document important events and preserve memories
- *emergency call and fall detection*—triggers an emergency call in situations where help is needed (either manually or by using a fall detection algorithm)
- *medication*—supports the administration of medication, e.g. medication reminders, errand lists for medication that has run out and security features such as warnings about maximum doses a day.
- *toolbox*—integrates a collection of easy-to-use tools to support daily life (for example, finding lost things)
- *state of health tracking*—includes assessment of current state of health and motivation to perform physical or mental tasks guided by the virtual personal assistant.

The user interface integrates a virtual personal assistant of human appearance that simplifies interaction with the system and allows interaction using natural speech (through speech recognition and text-to-speech).

By including interfaces for informal carers, DALIA allows them to stay in touch with their loved ones and care for them remotely, even if they do not live together. The easy-to-use interface provides the same look and feel across all devices (Fig. 2.1).

To build trust in the system and protect user privacy, DALIA integrates state-of-the-art encryption. Directed approval by the older adult allows other users to connect to DALIA. If the older adult has given them permission to access data, carers (informal carers and formal care and medical services) may provide support remotely, for example by adding appointments and reminders, managing medication or simply making a video call to have a chat. In turn, older adults can share their day-to-day experiences with carers through photos and messages, letting them take part in their life. If they want, older adults can give permission for automated, high-level notifications to be sent to carers in specific cases (for example, medication intake not confirmed).

- *DALIA* is an integrated but modular solution that is individually customisable to the needs of the individual.
- *DALIA* provides an interactive virtual personal assistant that knows its users and cares for them. It talks and understands natural speech so that all functionality can be easily accessed through its speech control interface, especially on the TV, tablet and laptop.
- The *DALIA* consortium involves real end users and care organisations that have already shown a great deal of interest in the solution. Additionally, the consortium has a powerful network of business partners with an interest in the solution and plans to sell it.
- No sensitive data is shared via cloud services outside *DALIA*. Proxy re-encryption is implemented, where the data host is only forwarding data and has no direct access to sensitive data. Sharing of data has to be explicitly approved by the primary end user for each module and each informal and formal carer.

2.2.3 Evaluation and Feedback

End users (older adults, informal carers and formal carers) were involved throughout the project in a multi-stage approach.

- The first step was the end-user elicitation phase. The elicitation plan used a step-by-step approach to optimise the data collection process.
- The second step was to formulate specifications for different use cases in three requirement workshops together with formal carers.
- In the third step focus groups were introduced: These groups of end users gave feedback on use cases, screens and designs. The aim was to collect information

Fig. 2.2 End user involvement in four countries

and find out if there were any obvious flaws in the requirements and design as early as possible before and during implementation. The procedure was iterated on a two-monthly basis, each time adding a new module (five focus groups held).

- The fourth step involved lab testing. These tests were designed to continuously evaluate DALIA's software. The aim was to test finished sections with end users during the implementation phase to improve the software while developing it further (five lab test sessions held).
- The last stage consisted of a series of field tests, where the finished prototype was tested in a realistic environment by each user for 3 weeks (Fig. 2.2).

2.2.4 Conclusion and Lessons Learnt

The *DALIA* project can be seen as a real success. At the end of the project a fully functional and evaluated prototype was available that served as a basis for commercial product development. One business partner—a start-up formed to develop the product further—planned to launch it within 2 years of the project closing with the support of additional consortium members.

Three factors contributed to the success of *DALIA* during set-up and execution:

- *Co-creation*: include all stakeholders in co-creation workshops with a focus on the end users as soon as possible and really listen to them. Have a good mixture between push (innovative new aspects you want to include in the system) and pull (features requested by end users). Prioritise requests clearly and bundle them into several, iterative, testable, well balanced system prototypes.
- *Architecture*: do not build everything from scratch. Where applicable use proven frameworks (such as Google services) that are stable and can be adapted to the specific needs of the project. This approach allows progress within these frameworks to be used in the project for free. Concentrate on the specific features of your project (encryption, adaptability) and differentiate yourself there. Design a KISS (Keep It Simple but Smart) system with lean and clear interfaces between the different modules and development responsibilities. There is not just one type of end user who will use your solution. Users have multiple needs, and these will change over time. Therefore, make sure the architecture supports a solution that is modular and can be customised. Furthermore, allow the solution to be configured remotely and provide help with appropriate data protection.
- *Collaboration*: ensure easy and frequent communication among all partners. This is essential for distributed project work. Regular online meetings and a clear task list facilitate information flow. When the project starts up, it is important that all the people involved meet face-to-face for a kick-off workshop and do some team-building activity. Then, during the project, agile development will allow fast prototyping with repeated feedback loops involving all

stakeholders. This helps to keep the project on track and minimise risk and deviation from the project goals.

As with every project, there were also some lessons learnt that may help to improve current and future projects:

- Start as soon as possible with working prototypes. Show and discuss them with all stakeholders very early in the process. Improve the prototypes iteratively and evaluate them, continuously involving all stakeholders.
- Involve the distribution partners at an early stage. Do not focus on end users only. The benefit (for the potential distribution partner) of being able to integrate the services easily into their existing business work flow is very important. Therefore do not just develop a lean end-user interface. Make integration, manageability and maintenance as easy as possible too.
- Using an avatar to hide the technical complexity of your solution and provide a natural way of interacting is a promising way forward. But facilitate direct human interaction too. Also, bear in mind that an avatar is only useful if it works reliably with dialects and different languages.

2.3 RelaxedCare—Unobtrusive Connection in Care Situations

HomeLab—Lucerne University of Applied Sciences and Arts, Trionic, Austrian Institute of Technology GmbH, 50 plus GmbH, New Design University, Mobili, Ralph Eichenberger Szenografie, Ibernex, soultank AG, Red Cross Switzerland Lucerne

2.3.1 Overview and Aims

Informal carers (ICs) always wonder how the person they care for is doing. Feelings of burden, stress and even burn-out are common because of all the different tasks the IC has to cope with. Currently, they reassure themselves about the assisted person's (AP) well-being through regular phone calls and visits, which can cause even more stress. However, most APs do not want to burden their ICs even more or disturb them in their busy daily life, and often perceive their own problems as minor. With that in mind, the system we developed has identified three major challenges:

1. To answer questions about how the AP is doing in an easily comprehensible and unobtrusive way by providing valuable information about his or her well-being.
2. To provide an easy way to stay connected with loved ones.

Fig. 2.3 User interface of the system consisting of a cube shaped object with a "Wellbeing state" indication (dynamically lit area in front). On *top* of the cube there are 3 lens-shaped tags for messaging. Next to the cube is a smartphone running the RelaxedCare-App. *Source* RelaxedCare Consortium, Ralph Eichenberger Szenografie

3. To combine these functions in a aesthetically pleasing lifestyle product. It must be fun to use for all generations, providing the potential to strengthen bonds within families and facilitate mutual caring.

2.3.2   Implementation

An aesthetically pleasing cube and a smart phone application (see Fig. 2.3) are the primary user interfaces. The two main tasks here are to inform the user of the well-being state (WBS) of the relative (AP) and to enable easy communication using simple signals. The cube-shaped object can output audio-visual signals and display reactions on physical message tags. An indication of the WBS of the remote relative (AP) is provided by the colour of the illuminated circular area on the front. There are three levels of WBS with clearly defined colours:

- green for WBS above average,
- orange for WBS average and
- purple for WBS below average.

The system can display a colour gradient, which can be useful for visualising trends. To protect privacy, the WBS indication is not updated more than three times a day, thus uncoupling it from current events at the place of the remote relative.

Messaging through the cube is done by physical tags. These are lens-shaped with a diameter of around 4 cm, and in tests, symbols representing a blue telephone receiver, a red heart and a green circle were attached. All tags and the symbols on them were printed in 3D. On the top side of the cube an area sensitive to theses tags can be found. If a tag is put onto the cube into this area it is recognised and initiates a tag specific functionality (e.g. "please call me" in case of the phone symbol). Further, to see that the command has been taken up by the cube, the tag initiates an audio-visual feedback—through a sound and blinking LEDs. The LEDs also indicate different communication states of the cube (e.g. "message sent", "message received" etc.). The feedback-sounds are pre-recorded and last less than five seconds. If a message received is not acknowledged by the user, such a sound may be repeated several times (e.g. every 10 min) with down-fading volume until it is finally discarded.

Fig. 2.4 System architecture. Sensors in the house of the AP are connected to the smart home server. The resulting wellbeing status (WBS) is shown to the IC via objects (cube, picture frame), or app. *Source* RelaxedCare Consortium, Ralph Eichenberger Szenografie

Although the system architecture allows for multiple cubes to be connected and synchronised in one apartment, they cannot serve as mobile devices. For users familiar with smart phones, a dedicated application has been developed, which, besides WBS and messaging, provides a history function. It allows users to obtain more detailed information on what has mainly influenced the WBS and to get an overview of the long-term trend (up to 6 months).

The algorithm infers the WBS based on sensor data. To detect the inhabitant's movement patterns and activities, at least three motion detectors, three contact switches and one bed sensor need to be installed in the home. All sensors are wireless and comply with the Enocean standard for data transmission. These sensors and the cubes are connected to the smart home server. It is based on an open, existing AAL platform supporting several (sensor) communication protocols (Fuxreiter et al. 2010). Located in the user's apartment/house, this server performs data fusion and executes the advanced pattern recognition algorithm for WBS analysis. Beyond that, it meets privacy requirements by ensuring that all sensor data is processed locally and only the WBS and messages leave the local network in an encrypted way. Figure 2.4 illustrates the system architecture for the use case where only the informal carer has a cube and sensors are placed in the older person's home. However, the system is designed to manage all modules (i.e. sensor and cube installations) on both sides.

2.3.3   Evaluation and Feedback

End users (i.e. ICs and APs) were involved in all project stages, from the elicitation of the user needs up to the planned commercialisation stage. A strong commitment

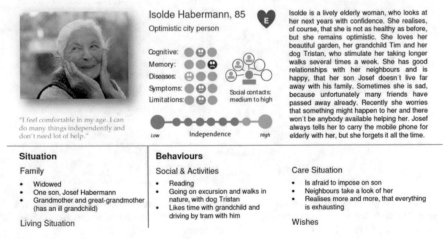

Fig. 2.5 Excerpt of a persona (AP) derived from initial requirements engineering. *Source* Relaxed Care Consortium, Soultank AG

Table 2.1 Characteristics of ICs

IC	1	2
Age	60	58
Gender	M	F
Family situation	Divorced Two children Already a grandfather	Married to Christian for 20 years Two children and one grandchild
Living situation	Lives in a company flat in a city Doesn't know his neighbours Daily routine is enough work for him	Lives in a house in the suburbs of a city Her father lives on the first floor of the same house Has lunch with the whole family and her father
Care situation	Two times a week he visits his mother He wants professional help for his mother	Cares for her aunt, as she feels alone sometimes Lives in the same house with her father-in-law and helps him several times a day
Technology usage	Technophobe Has an old computer for email	Technologically skilled If there is a problem, her husband helps, he works for an IT company
Communication	Uses only the landline phone If he has to communicate, he prefers face to face contact As his son lives abroad, they stay in contact by email	Uses a smart phone and iPad Can't afford the latest gadgets

of the project team to user-centred design resulted in applying ISO 9241-210 throughout the whole development life cycle. Furthermore, the 'user-inspired innovation process' was applied consisting of the following steps: ignite, perceive, collect, decode, assemble, experiment and merge. This process incorporates knowledge and methods from different disciplines of academic research, design and industrial practice (Fig. 2.5).

End-user involvement in Austria and Switzerland involved 207 participants, and the following methods were employed in the field of design research: (1) personas (assumption), (2) show and tell method, (3) cultural probes, (4) design workshop, (5) questionnaires, (6) focus groups, (7) discussion and contextual inquiry interviews. These methods are described in Dittenberger and Koscher (2015) and Dittenberger et al. (2016) (Tables 2.1 and 2.2).

Qualitative tests with end users have shown there is a need for a socio-technical system such as RelaxedCare. In preliminary trials a cube shape was identified as being sufficiently neutral to fit into any interior. The option of exchanging simple messages in the form of signals ensured that social contact by telephone was not impeded by the system and opened up the option to participants of sending short, non-urgent messages that did not require an immediate answer. The benefits of this are taken for granted by text message and instant messenger users, but are rarely

Table 2.2 Characteristics of APs

AP	1	2	3
Age	85	83	82
Gender	F	M	F
Family situation	Widowed One son Grandmother and great-grandmother	Widowed Two children and two grandchildren	Widowed No children Supported by niece
Living situation	Community housing, lives alone Friendly relationship with neighbours Loves her garden Adores her dog	Lives in his own flat on the first floor of his daughter-in-law and grandchildren's family house Has a good relationship with his family and the neighbours	Lives in the suburbs in an apartment for elderly people Half an hour away from her niece
Care situation	Is afraid to impose on son Neighbours look in on her Finds everything more and more exhausting	His daughter in law looks after him quite often during the day Grandchildren visit him for a chat a few times a week Eats with the whole family once a day	Never wanted to live in a nursing home Has visits from a professional carer two times a week Gets lunch daily
Technology usage	Technophobe Has landline phone Has mobile phone for the elderly, but doesn't use it	No affinity with technology A few times a week he listens to a midday talk show on the radio Likes to watch TV	Landline-phone Emergency-system
Communication	Likes face-to-face contact Sometimes telephone calls Visit from son two times a week	Uses the land line phone Has a lot of visitors Writes postcards	Loves visits from her niece Likes observing her neighbours Chats with neighbours a few times a day

accessible to those who are less information and communications technology (ICT)/mobile literate. Furthermore, the haptic nature of communication through tags was appreciated. The colour coding of the tags and the tangible symbols attached to them allowed people with low vision to recognise them, and a grandchild reported that handling tags was fun.

To allow prototypes to be developed rapidly, all technical system modules (server, cube, router) were kept separate. In addition, the dimensions of the cube were limited by the footprint of the electronics inside. Consequently, test users often suggested reducing the size of the cube and the number of separate components. From a technical point of view, this should be feasible for future products.

Although the system is different from medical devices or time-critical emergency and telecare solutions, it must be very reliable for the user to accept it. Otherwise the goal of bringing the informal carer peace-of-mind is defeated. The system becomes an additional burden if you have to wonder whether it's working. Therefore, if the connection between the AP and the IC is lost, this is indicated by a slowly moving white dot where the WBS would normally be. To avoid any suspicion that the system might have an adverse effect on the local Internet infrastructure, all systems were equipped with their own Internet access through a GSM router. In field trials, the connection through GSM routers proved not to be as reliable as expected.

Based on further end-user feedback, sound output could also be improved. Even after the sound output had been reduced to a minimum—earlier versions of the cube played continuous, soft, ambient sounds—people wanted to configure the sound output themselves. While this is technically feasible, creating a user interface that allows less computer-literate users to configure features themselves is difficult. To overcome configuration and installation obstacles, the IC was identified as the target customer group for this system. In future marketing, it might be best not to brand the system as an assistive system for the elderly but instead as a lifestyle product with the communication features in the foreground. Such a product could offer extra features, for example upgrades with additional sensors and functionalities, when changes in the AP require them. For the system to be accepted, it is important not only to upgrade it on the AP side, but also in a reciprocal fashion on the IC side. Running a system with sensors on both sides not only quashes the argument that a system such as this is only for surveillance of the AP, but also addresses the AP's legitimate interest in the well-being of their offspring (IC). Potential positive side effects from this principle of reciprocity are better acceptance of the system by the AP and because of the reassurance provided by a more ICT-literate relative trusting the system. Finally, the IC will be motivated out of self-interest to keep use of sensors to an absolute minimum in both households.

Additional information about the system developed is available in Morandell et al. (2016), Redel et al. (2015) and Morandell et al. (2013).

2.3.4 Conclusion and Lessons Learned

Ensuring stakeholder involvement in all development phases, rapid prototyping and testing with end users is an effective way of developing a product. It's worth remembering that such products are rarely purchased by less ICT-literate elderly people because of legitimate concerns about complex installation and configuration. Furthermore, the product should not be pitched as an assistive system for impaired people. Where possible, it should follow universal design rules and benefit as many different users as possible. In the context of this project, the IC is the main target customer group. This group appreciates products that can be configured by the user, and such products will therefore be more successful than static designs.

The approach of designing a reciprocal system, with sensors detecting movements and activities on both sides (those of the AP and the IC), is innovative and not widely used. It creates an equal relationship between the IC and the AP, increasing acceptance among the latter compared to unidirectional monitoring systems only letting IC observing their AP.

Products that help less ICT literate elderly people—beyond medical devices, telecare and emergency systems—are a seldom exploited niche in the market, which, in our view, would repay further exploration.

2.4 Confidence—Mobility Safeguarding Assistance Service with Community Functionality for People with Dementia

iHomeLab—Lucerne University of Applied Sciences and Arts, Salzburg Research Forschungsgesellschaft m.b.H. Raiffeisenlandesbank Kaernten reg Gen. m.b.H., ilogs mobile software GmbH, Presence displays bv. Ralph Eichenberger

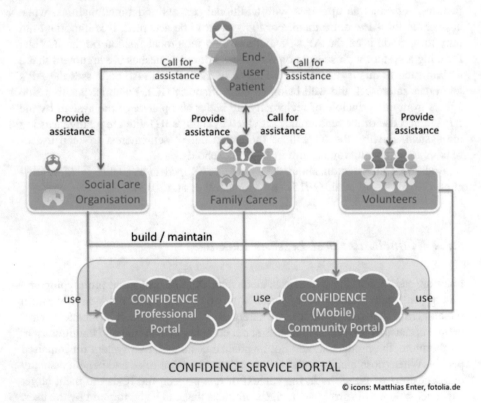

© icons: Matthias Enter, fotolia.de

Fig. 2.6 Community support in Confidence. *Source* Confidence consortium

Szenografie, Hilfswerk Salzburg, Kanton Zug, Ana Asalan International Foundation, Swisscom Participations Ltd.

2.4.1 Overview and Aims

As the population ages, dementia is a growing problem. Dementia is an umbrella term for a group of diseases characterised by a progressive and irreversible decline in brain function with symptoms such as memory loss, disorientation, cognitive decline and inappropriate social behaviour. It is overwhelming not only for the people directly affected, but also for their families and carers. Dementia is one of the major causes of disability and dependency among older people worldwide.

Thus, Confidence aims to provide adaptable mobility and safeguarding assistance services to people who suffer from mild to moderate forms of dementia by combining assistive technologies with personal help. On the one hand, Confidence aims to support the affected people (primary end users) to be more independent and active for as long as possible; on the other hand, it seeks to decrease the burden on their carers (secondary end users). The project addresses the following issues: personal contact between the primary and secondary end users when help is needed;

Fig. 2.7 Primary end user assistance app. *Source* Confidence consortium

support in case of an emergency; assistance with spatial and/or temporal disorientation; and support of caring communities (Fig. 2.6).

Each issue is dealt with by a different Confidence service:

- *Voice and localisation service* allows the primary end user to request assistance
  from carers, who can—if consent has previously been provided—locate the
  person in need on an electronic map, see the immediate surroundings over
  the built-in smart phone cams and give instructions on where to go.
- *Video service* allows the primary end user to see the face of the carer, thus
  creating a feeling of confidence and security.
- *Mobile community service* enables carers to be mobile themselves while giving
  instructions to the person who is disoriented.

These three services are provided via two mobile applications and a web portal.
The user interfaces were developed in close cooperation with end users. The
implementation for the primary end user was done in the so called *Confidence
Assistant* App which aims to support primary end users in their daily life, offering
the following functions (Fig. 2.7):

- *Emergency*—Users press *SOS* if they need help urgently. In the background the
  system informs the responsible carer.
- *Assistance*—Users need help and would like to talk to somebody personally.
  Confidence enables a video or voice connection with the responsible carer when
  users press *Support*.
- *Daily schedule/reminders*—Pressing *Calendar* shows the users' tasks and
  appointments of the day. Further entries can be added. Reminders for entries
  appear automatically and are read aloud at pre-set times.
- *Navigation*—If primary end users need support on their way home, they click
  Maps and are able to see a map showing the route and distance from their
  current position to their home address.
- *Environmental service*—If users press *Weather*, the current or short-term
  (adapted to the appointment in the calendar) weather conditions are shown with
  tips on suitable clothing.

The counterpart to the *Confidence Assistant* is the *Confidence Helper* application
for informal carers, which provides:

- *Alert management*—clicking *Alerts* shows all open alerts from one or more
  primary end users and their status
- *Call primary end user*—clicking *Support* means the connected primary end user
  is called
- *Localisation*—if there is an open alert, a map with the current position of the
  concerned primary end user is shown
- *Calendar management*—pressing *Calendar* enables users to view and manage
  their own and the primary end user's calendar entries. New reminders can be set
  up.

The web portal enlarges the Confidence Helper application. Most features (for example, localisation in case of an emergency, alert overview, calendar management and so on) are also available on the web portal. Additionally, secondary end users are enabled to administer their own as well as the connected primary end user's data on the web portal and to apply extended features such as geo fencing to the map.

2.4.2 Evaluation and Feedback

The project followed the human-centred design standard for interactive systems (ISO 134079). Therefore, end users were involved in all phases of the development process: (1) requirements analysis, (2) system design, (3) implementation and (4) field trials. The services were developed in two subsequent iterations.

After establishing the requirements, end users were involved in the design phase. Two so called acceptance tests were conducted to determine the usability and functionality of the system. Both tests were conducted as workshops, meaning that potential end users tested and evaluated prototypes in an open discussion. The first acceptance test focused on evaluating the design. Features were presented in different design variants. The user interface was evaluated by primary and secondary end users in Switzerland and Austria. After the feedback was evaluated, the advantages of the different prototypes were combined in one user interface. For example, user feedback on contrast, button size, fonts and colours were taken into account to develop the next version of the user interface. The second acceptance test was conducted in Switzerland, Austria and Romania and focused on evaluating the features of the mobile services. End-user feedback was again analysed and summarised to improve implementation further.

Once the acceptance tests and the initial development cycle were over, the first field trial was carried out in Switzerland, Austria and Romania. One hundred and eighty end users (126 elderly people and 54 formal and informal carers) tested Confidence over six weeks in their daily living situation. In order to involve suitable end users, the MMSE (Mini Mental State Examination method to diagnose dementia patients) was used to estimate the stage of dementia of the primary end users. Family members and friends tested the Confidence Helper in field trials by using their own smart device or a unit on loan. Employees of the end-user partners were trained to support the end users during the field trials (first-level support, train-the-trainer concept). The trainers also documented the trials at least once a week using a standardised protocol. At the end of the trials, they filled in questionnaires with the end users. In parallel, an evaluation guideline for the implemented service was prepared for the second field trial to measure the effects of usage (changes in quality of life, safety and mobility, autonomy and independence, daily assistance and usability).

This guideline was used to evaluate the second field trial where 168 end users (elderly people, volunteers, formal and informal carers) used Confidence over 6 weeks in their everyday life. During this last field trial, the Confidence application was generally accepted and considered useful for people with mild to moderate cognitive impairment. However, at the beginning of the field trial, almost two-thirds of the end users were reluctant to use the app. This was because they weren't acquainted with the system and weren't aware of the benefits of using the services provided by the Confidence app. They perceived themselves as being unable to learn how to use the various services. Both hurdles were diminished by sustained training with a human trainer. Despite frequent technical problems during the first phase of the trial, more than 60 % of the users were 'highly satisfied' with their use of the app at the end of the field trial. Reasons mentioned for this high acceptance rate were that the app provided an opportunity to communicate more easily with relatives, helped them to plan their daily activities properly and provided useful support through the 'Weather', 'Help' and 'Calendar' features. The least preferred services were 'SOS' and 'Map'. About 17 % of participants rejected the platform outright, believing they couldn't learn the necessary new skills or claiming that the services weren't useful in their daily lives, particularly at their age. Such attitudes were also found among a considerable proportion of the informal carers. Being elderly themselves and in the completely new and exhausting situation of having to care for their spouse, they didn't want to be confronted with a new ICT-based service at that time.

2.4.3   Conclusion and Lessons Learned

The above outcomes should be taken into careful consideration because services often malfunctioned due to poor network connections. At the beginning of the evaluation, there was some scepticism about this application. However, attitudes changed during the evaluation, mostly for the better. Broadly, user reactions can be divided into two groups: those related to technical problems and those related to the application services themselves.

It is therefore important to make sure that the application tested by end users is very stable and does not display messages or texts that are not intended for them. If the application relies on a network connection, it's important to ensure the testers have a sufficiently good connection where they live. This is especially important for tests in rural areas or abroad.

It's also useful to have a 'killer feature' that is greatly appreciated by end users. In Confidence, the 'Weather' service—though very simple and unspectacular—was one of most used services and contributed to Confidence being accepted. This service also provided clothing tips—reminding people to wear sturdy shoes in snowy weather or to take an umbrella when rain was forecast, for example—and almost everyone appreciated it. We believe it is important to make a product useful in day-to-day life and not just in emergency situations. It's also best not to focus on one very specific user group (such as 'people with dementia') but to include other

Fig. 2.8 AAL award 2014 ceremony. *Source* Confidence consortium

user groups as well. Our aim should be to develop solutions and create products that could potentially be useful for many people of different ages and backgrounds.

The Confidence project greatly benefited from the complementary expertise and experience that the partners brought in. The consortium consisted of 10 organisations from four European countries, including end-user organisations, small and medium sized companies (SME), large industrial companies and research organisations. Therefore we were able to rely on the expertise of technology providers, user interface experts, infrastructure providers, business developers and social care organisations. The end-user organisations in the three trial countries enabled user integration which was essential for creating and testing the Confidence services. Coordinating such a large and diverse consortium is challenging, but this was handled very well by the Austrian coordinator. All of this is proven by the fact that Confidence received the AAL award in 2014 (Fig. 2.8) and that the idea is pursued further by one of the commercial partners of the consortium with the aim to bring a product to the market.

2.5 iWalkActive—The Active Walker for Active People

iHomeLab—Lucerne University of Applied Sciences and Arts, Trionic, Austrian Institute of Technology GmbH, Careguide GmbH, Ith-Icoserve, Trikon AG, Geo7 AG, Kanton Zug.

2.5.1 Overview and Aims

2.5.1.1 Why iWalkActive?

Rollators have become very common mobility aids, and in countries such as Germany and Sweden about 4 % of the population uses a rollator. However, a significant problem with existing rollators is that people in need of support often hesitate or refuse to use them. When trying to encourage someone who would clearly benefit from using a rollator to do so, you often hear: 'No, I'm not that old!' or 'Never, I'm not that sick!'

Instead of seeing opportunities to stay physically active, a lot of people tend to associate rollators with being old, ill and 'handicapped'. Rather than blending in with other modern products, rollators stick out as being something for 'the old', creating stigma.

A few innovative rollator features have been developed in recent years, for example user-friendly folding. But the performance, comfort level and driving characteristics that a standard rollator offers when used for walking still leaves a lot of room for improvements. Problems often arise outdoors when the user needs to overcome a physical obstacle, such as a kerb or cobblestone, or go over uneven ground, such as gravel, grass, sand or snow. When compared with vehicles such as bicycles, baby strollers and cars, the rollator has not yet entered the new millennium.

The aim of iWalkActive was to offer people a highly innovative and attractive walker with additional services that greatly improve the user's mobility in an enjoyable and motivating way. iWalkActive is designed to encourage/enable hobbies and physical activities that would be impossible or very difficult to pursue with a traditional rollator.

2.5.1.2 Problems Identified by the End Users

End user requirements were collected through an online survey, six focus groups that included end users and potential end users as well as healthcare professionals, a Delphi Study with five rollator and healthcare experts and finally two presentations followed by discussions among professionals who specialise in improving the living situation of elderly people. Through the involvement of end users and professionals, the following problems with rollators were identified or designated high priority:

1. difficult to pass obstacles like kerbs and rocks
2. difficult to walk on uneven ground, for example cobblestones and gravel
3. difficult to use public transport (car, bus, train)
4. difficult to walk downhill (rollator rolling away from the user causing falls)
5. pain or discomfort because of the rollator shaking on uneven ground
6. difficult to walk uphill.

The following areas of use were rated most important by end users as daily living and leisure activities that could possibly be supported by ICT-enhanced services:

- Shopping
- Traveling/sightseeing
- Hiking
- Fitness walking

2.5.2 Implementation

2.5.2.1 E-drive

The integrated e-drive consists of two hub motors in the rear wheels. One of the main challenges was to control the motors, since the end users naturally wanted the Active Walker to behave they way they expected it to behave. This is not very different from a standard rollator, which should allow users to walk in the direction they want at the speed they want. The user should be able to steer the Active Walker as intuitively as possible without noticing any difference when walking straight or up or downhill.

By implementing force control in the rollator handles a very intuitive human-machine interface was created. Users feel constant inertia when pushing the walker, irrespective of walker load. If users want to steer right or left, they intuitively apply more force to one hand grip, which means that the adjacent motor rotates faster (Fig. 2.9).

The e-drive and its controlling mechanism are composed of several key components:

1. force sensors fully integrated into the right and left hand grips
2. control unit offering two e-drive modes: a gentler and a more direct one
3. battery (rechargeable)
4. two independent rear motor wheels
5. push-button control for switching between the two modes of the e-drive
6. mechanical brake handles, integrated with the rear motor wheels

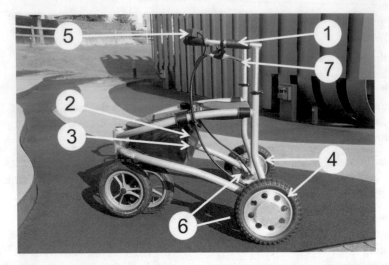

Fig. 2.9 The Active Walker with the major components of the e-Drive indicated. *Source* iWalkActive Consortium, iHomeLab

7. mechanical brake levers, electrically connected to the control system.

2.5.2.2 Localisation

For navigation purposes, the position of the user needs to be known. This is done with the help of wireless localisation technology. While outdoors, the global positioning system (GPS) is omnipresent, nothing comparable is available indoors. We used two technologies for indoor localisation: WLAN and iBeacons.

2.5.2.3 Seamless Transition

For the localisation service we faced the challenge of selecting the best source for localisation and for a seamless transition between outdoors and indoors. The method applied determines if users are indoors or outdoors based on available location sources.

2.5.2.4 Open Data Integration

During the project it became useful to gather geo-information from open data providers, for example to list walking routes or public toilets. The iWalkActive Open Data App showcases data that can be fetched from open data providers for free. The data is integrated into a Google maps view and can easily be used to help users in finding a hiking route or different points of interest (tourist attractions, restaurants, toilets etc.).

2.5.2.5 Navigation

Using seamless localisation technologies as described above, the Active Walker knows the users' position and whether he or she is moving indoors or outdoors. To give directions and guide the user along the right path, localisation is accompanied by a navigation system.

The navigation client app receives most of the data, such as calculated routes and geo codes, from the Geo Information System or short: GIS-server. Each client sends a routing request to the Network Analyst Extension on the ArcGIS Server. The client then receives a complete optimised route with directions. With the positioning information, it is now possible to guide the user to the destination. The navigation in iWalkActive requires the calculated route to avoid obstacles that cannot be passed by a user with a walker (no stairs are allowed, for example).

A further feature is the 'closest facility' function. With a separate button, users can select the shortest rollator-friendly route from their current position to the nearest barrier-free toilet. Of course, other facilities can be implemented, such as navigation to the nearest shopping centre.

2.5.3   Evaluation and Feedback

Using different tests, we ensured that all components in the iWalkActive system worked safely. We divided the tests into lab testsperformed internally by our research staff and field trials performed by potential end users.

2.5.3.1   Lab Tests

The scenarios investigated in the lab tests guaranteed the smooth functioning of all the integrated technologies and implemented software services. The emerging prototypes were tested at least twice at different stages of development. Where external people were involved, the lab tests were undertaken using the model to test user acceptance developed by Mollenkopf and applying some of the insights from mobility studies in which she was involved (Tacken et al. 2005).

Test results showed that most users found the motor in the active walker to be helpful when walking on rough and level surfaces outdoors. Many elderly people said that the motor was helpful when walking on smooth outdoor surfaces, but even more helpful on rough ground, and especially when going uphill or downhill with a load (for example, heavy groceries in the rollator bag). In general, they thought it could be of help for shopping trips, sports training and longer walks—especially on rough ground—and on holidays that would not even have been planned without such a motorised walking frame. Training aspects were mainly mentioned by men, while women liked the possibility of going for longer walks, travelling and making shopping easier.

2.5.3.2 User Field Trials

Field trials involving end users took place in different locations and under real conditions: Lucerne University of Applied Sciences and Arts in Horw, Switzerland; Austrian Institute of Technology in Wiener Neustadt, Austria; Sveriges Pensionärsvörbund in Uppsala, Sweden.

The rating of the e-drive support in different test settings showed that most subjects were very satisfied with it. The best ratings were given for rough ground (with or without a load). Ratings were also very good for how the rollator handled on bends and the motorised braking system. Only in the scenario 'going round a bend downhill with a load' did some of the test candidates say the support was too powerful.

For navigation functionality, the user interface is crucial. Tests have shown that it could be improved, especially with regard to intuitiveness (i.e. making it immediately clear to the user which control element stands for which function). Apps on standard tablets are affected by outdoor lighting conditions. Therefore, visibility, legibility and contrast need to be enhanced. The navigation function was not rated very highly. The benefits were regarded as 'medium' in the situation our users found themselves in. However, this may be due to the fact that the generation that took part in the study is not (yet) used to these kinds of devices. Additionally, it may not have been clear to the target group during the tests in which kinds of situations the navigation function could be beneficial. To improve this, test scenarios may have to be adapted and further, longer field trials may be necessary to get more insight into how the navigation features operate in daily life.

Among candidates using the rollator at home for several weeks, the outdoor capabilities of the Active Walker were greatly appreciated. One person, a multiple-sclerosis patient, showed great interest in buying the prototype. He lived out of town, and his wife told us that he liked the Active Walker because it allowed him to take a daily walk in the hills again, which he had not been able to do for a year before receiving the Active Walker in the trial.

2.5.4 Conclusion and Lessons Learned

From our point of view the project, which won the European AAL Award 2013, is very successful. We have shown that an e-drive powered rollator with seamless indoor and outdoor navigation is feasible. The consortium worked very well together, but there is room to improve the efficiency in future projects now that the partners know each other. A debriefing meeting confirmed that face-to-face meetings are important. The partners said they would meet more often in future, not only in plenum but also in smaller working groups focused on technological issues, business development, and planning and organising the end user involvement.

The involvement of end users and/or potential customers is crucial. Therefore, contact with end users should happen as often as possible. But this is also a challenge: to get useful feedback, you need to present working prototypes. As prototypes for a hardware-driven project such as iWalkActive can become quite complex, relying on different partners for the frame, brakes, motor, motor controller, sensors, mobile apps and so on, manufacturing, integrating, testing and delivering enough walkers is challenging. The costs of creating a hand-crafted prototype, for which many of the production process steps are still done manually and for the first time, should not be underestimated (in the case of iWalkActive, the motors for the e-drive powered rollator were a completely new development). In our project, the initial budget was enough to produce three complete walkers (others only had part of the functionality). In a similar future project, we would dedicate more manpower and material resources to that.

Another aspect is the business side. Although we developed several business models, we did it rather late. The business model should be on the agenda from the very beginning, and you have to deal with the fact that the business model evolves as the product does. Our three business partners decided to develop further a distributed business model. The provider of the geo information modules, which handle localisation and navigation, founded a start-up to promote the solution. The rollator and e-drive provider decided to develop a solution in which the e-drive is sold and delivered to the Swedish rollator company that integrates it and sells it to its customers. The biggest challenge will be to bring the price down to a level that end users will tolerate in order to make a business out of it. However, recently other projects and even manufacturers have also come up with the idea of e-drive rollators, creating prototypes and even first products. As awareness grows and the market expands, new opportunities may open up for our Active Walker too.

Acknowledgments The four projects Confidence, Dalia, iWalkActive and RelaxedCare were co-funded by the AAL Joint Programme and the following National Authorities and R&D programs in Switzerland (SERI), Austria (bmvit, FFG, Programm benefit), Slovenia (MIZS), Spain (National Institute of Health Carlos III, and Ministerio de Industria, Energía y Turismo (MINETUR)), Belgium (IWT), NL (ZonMw), Rumänien (UEFISCDI), Schweden (VINNOVA).

References

Dittenberger S, Koscher A, Morandell M et al (2016) RelaxedCare: an iterative user involvement project. In: Design and emotion conference. Amsterdam, 27–30 Sep 2016

Dittenberger S, Koscher A (2015) How much design does research need: an inquiry of the synergetic potential of methods of social and design research. In: International conference on engineering design. Politecnico di Milano, Milan, 27–31 July 2015

Fuxreiter T, Mayer C, Hanke S et al (2010) A modular platform for event recognition in smart homes. In: 2th IEEE international conference on e-health networking applications and services (Health.com), Lyon, 1–3 July 2010, pp 1–6. doi:10.1109/HEALTH.2010.5556587

Morandell M, Dittenberger S, Koscher A et al (2016) The simpler the better: How the user-inspired innovation process (UIIP) improved the development of RelaxedCare–the entirely new way of communicating and caring. In: Marcus A (ed) Design, User Experience, and Usability: Technological Contexts. 5th international conference, DUXU 2016, held as part of HCI international 2016, Toronto, Canada, July 17–22, 2016, Proceedings, Part III. Springer International Publishing, pp 382–391

Morandell M, Mayer C, Sili M et al (2013) RelaxedCare—Unobtrusive connection in care situations. In: Proceedings of the AAL Forum. Norrköping, 24–26 Sep 2013

Redel B, Uhr MBF, Morandell M et al (2015) RelaxedCare—Connecting people in care situations: user involvement to collect informal caregivers needs, AAATE Conference 2015, Budapest (Hungary). Will be published in the proceedings of the 13th AAATE Conference, Budapest (Hungary)

Tacken M, Marcellini F, Mollenkopf H et al (2005) Use and acceptance of new technology by older people. Findings of the international MOBILATE survey: 'Enhancing mobility in later life'. Gerontechnology 3(3):126–137. doi:10.4017/gt.2005.03.03.002.00

Chapter 3
Using Gaze Control for Communication and Environment Control: How to Find a Good Position and Start Working

Maxine Saborowski, Claudia Nuß, Anna Lena Grans and Ingrid Kollak

Abstract The research project 'EyeTrack4all', based at Alice Salomon University of Applied Sciences in Berlin, showed that using gaze control to operate a communication aid or computer provides vital access to self-determination and participation for people who are permanently or temporarily unable to move and express themselves. In addition, a person with severely impaired motor function can use gaze control to operate various domestic devices. This chapter explains how gaze control works, what the prerequisites are for using gaze control, what to watch out for in tests and how a user achieves good positioning.

Keywords Alternative communication · Augmentative communication · Eye track · Gaze control

Picture a woman with amyotrophic lateral sclerosis (ALS). She can no longer move independently, she breathes with a ventilator and it's hard for her to express herself clearly. However, she'd like to carry on living with her partner and children. She'd also like to communicate with her carers and keep using a computer for as long as possible to write emails, watch films and exchange with other people. Furthermore, she wants to remain as independent as possible, by for example turning the light on and off, using the sound system and changing her position.

A second example might be a small boy in a wheelchair, whose muscles for movement and speech have been impaired since birth by severe spasticity. Ideally, he should be able to go to school and take part in lessons with the other children, study by himself, play with the other children in the breaks, and download music, surf the Internet and play computer games in his spare time.

Electronic supplementary material The online version of this chapter (doi:10.1007/978-3-319-42890-1_3) contains supplementary material, which is available to authorized users.

M. Saborowski · C. Nuß · A.L. Grans · I. Kollak (✉)
Alice Salomon University of Applied Sciences Berlin, EyeTrack4all (D), Berlin, Germany
e-mail: kollak@ash-berlin.eu

I. Kollak (ed.), *Safe at Home with Assistive Technology*,
DOI 10.1007/978-3-319-42890-1_3

A (voice) computer could be of great help to both. However, this would require the device to be operated by the eyes rather than by hand. Many people already use gaze control, but many more could benefit from it if more information were available and more professional help for potential users to show them how to use gaze control.

This chapter looks at who can use gaze control, how gaze control works, how to position users correctly in front of a gaze control device, which programmes and interfaces are available, what to look out for when servicing hardware and software, and what environmental control can do. It also gives practical everyday tips for carers and for advanced users. The chapter closes with suggested training materials.

3.1   Who Can Use Gaze Control?

Gaze control is suitable for people whose motor functions are severely restricted because of a birth defect, an illness or an injury, and who cannot communicate verbally or only to a limited extent. With gaze control, these people can use a communication aid or a computer.

This user group has very diverse limitations and capabilities. It ranges from young children with severe multiple disabilities, through people who have had traumatic brain injuries to elderly people with degenerative illnesses (for example MDS). What the users have in common is severe motor impairment. With gaze control, they can learn to communicate and maintain social contact. Furthermore, it's possible to use all the functions of a PC with gaze control, and so they can use email and the Internet to participate in family and social life. If environmental control is installed on the communication aid or computer, people can also control technical devices with their eyes, open and shut doors and shutters, and regulate lighting, and thus organise their surroundings despite severe motor impairment.

3.2   Why Is Communication Important?

There are at least two principles that justify a person's claim for communication options and support.

1. According to the International Classification of Functioning, Disability and Health (ICF) from the World Health Organisation (WHO) every person has a right to participate, and health promotion measures should deliver this vital asset. Communication enables participation and activity (self-determination), and for this reason appropriate steps to support communication should be taken as part of health promotion measures.
2. Article 21 of the UN Convention on the rights of persons with disabilities sets out the 'right to freedom of expression and opinion, and access to information'.

Fig. 3.1 User with his nurse
and gaze-controlled
communication aid

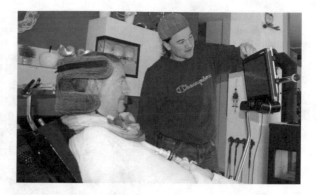

It then describes the measures necessary to support people with disabilities: 'States Parties shall take all appropriate measures to ensure that persons with disabilities can exercise the right to freedom of expression and opinion, including the freedom to seek, receive and impart information and ideas on an equal basis with others and through all forms of communication of their choice, as defined in article 2 of the present Convention, including by (Fig. 3.1):

'a) Providing information intended for the general public to persons with disabilities in accessible formats and technologies appropriate to different kinds of disabilities in a timely manner and without additional cost;
'b) Accepting and facilitating the use of sign languages, Braille, augmentative and alternative communication, and all other accessible means, modes and formats of communication of their choice by persons with disabilities in official interactions; […]'.

3.3   How Gaze Control Works

A gaze control system comprises infra-red light sources and one or more cameras installed underneath the screen of a communication aid or computer (Majaranta et al. 2012). The infra-red light illuminates the face, and the cameras film the head, the eyes and their line of vision. Infra-red light is used because it doesn't dazzle the user. Interference can arise if there is a lot of infra-red light in the ambient light, for example in direct sunlight. The infra-red light creates two small reflection points on the tear film of the eyes underneath the pupils. The cameras film the face with the pupils as well as the reflection points and transmit the pictures to the software. The software uses the data to determine where the person is currently looking at. In this way, the effect of head movements can be eliminated, so that only eye movements are taken into account in control functions. The advantage of this is that people

Fig. 3.2 Section of a camera
picture with the two reflection
points on the pupil and
interference from reflections
on the lenses of the glasses

who, for example, suffer from spasticity, can move their heads to some extent
without the gaze control cameras losing the eyes or confusing head movements with
eye movements (Fig. 3.2).

For people who wear glasses interference can arise from reflections on the lenses
of their glasses. These must not cover the reflection points. Many devices can
show the reflection points and possible interferences in the camera picture. It's then
possible to change the lighting conditions, or to try different glasses or contact
lenses.

3.4 Gaze Control as an Access Method

When a computer or communication aid is operated by gaze control, eye movements replace the mouse to some extent. The cursor is where the person is looking,
and it moves where the gaze moves. As the human eye moves a great deal
involuntarily without us realising it (saccade), mouse movements are based on the
main directions of movement and are reduced accordingly.

To select a field, the user looks at it for a certain, individually adjustable period
of time ('dwell mode'). Alternatively, selection can be done by blinking or by
pushing a button installed for this purpose, which is only used to confirm selection
of the field being looked at. If there is sufficient motor function, this combination is
a good solution as it gives the eye muscles a rest.

3.5   What Are the Prerequisites for Using Gaze Control?

Before gaze control is purchased, the person's current situation is assessed. The situation and the person's existing wishes for improvement are ascertained to create an overview of the communication situation, the person's resources and the resources available in their surroundings. Who wants gaze control? Who will be able to monitor the device later? To improve the communication situation, it's important to ascertain the wishes of the person and the people they most often communicate with.

It's helpful to ask at the planning stage who can perform certain tasks, such as looking after the device. It's good to clarify which alternative approaches to helping with communication have already been tried. Head control or using a button as access method may be alternatives (for the criteria for selecting a suitable solution see Blackstone and Pressman 2016 and Fager et al. 2012).

Gaze control can also be used by people with low vision. The basic requirements are that it must be possible to move at least one eye with control and that the camera must be able to recognise this eye.

3.6   How to Achieve Good Positioning

To use gaze control well, the user must be correctly positioned. It's helpful to check as soon as possible which positioning is most comfortable and functional and if any aids are needed for that positioning. It's important to position the head in a stable and supported way and to have a comfortable and relaxed body posture. In the beginning, it feels strange to use the eye muscles so much and in such a deliberate way. To find a good position, it's best to try out all the possibilities on an individual basis (see Andres 1996).

The screen should be about arm's length away from the face and parallel to it. The top edge of the screen should be a little above eye level. That is the recommended height for optimal use. If control with the gaze is working very well and there are requests, for example, to be able to see other people better over the device, a lower height can be tried out. For wheelchair users who control their chair themselves, the device must be mounted at suitable lower height so it doesn't impede the user's view (Figs. 3.3 and 3.4).

Fig. 3.3 Distance about 55–70 cm

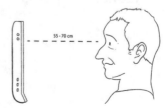

Fig. 3.4 Screen parallel to
the longitudinal axis

Fig. 3.5 Example of a
tabletop stand

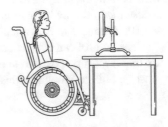

Fig. 3.6 Example of a
wheelchair mount

The camera must be able to see the person's eyes. This can be tested by checking the camera picture or by getting another person to look from where the camera is positioned. If the device is mounted very low down, it may be difficult for the camera to recognise the eyes, as lowered eyelids may hide the pupils. If the screen is fixed higher, this can open the eyes more.

A device can be set up in different ways (Figs. 3.5, 3.6 and 3.7).

Fig. 3.7 Example of a
mobile stand

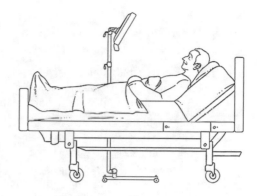

3.7 What to Watch Out for in Tests

When gaze control is being tried out for the first time, only a few people should be
present who know the person well and with whom they communicate a lot. It's
important to create a pleasant atmosphere. The device is being tested, not the
person. The main question is whether the person can control something with their
eyes or can learn to do so. The communication interfaces on offer are fun and
adapted to the person's interests. The sitting can be filmed so that other people can
get an idea of how the test went. The test sitting should be evaluated calmly by all
those involved.

If a device is purchased, the first units should be carefully planned with relatives,
carers and therapists, and the device's use should be discussed with all communication partners: which programmes and interfaces is it best to start with? The
person's capabilities and interest are the deciding factor. Games that familiarise
people with gaze control, such as *LookToLearn* or *SensoryEyeFX*, are useful for
learning how to operate the device alone. A simple interface with a selection of
photos or actions can also be designed.

Enjoyable and interesting tasks are helpful for getting the person gradually used
to controlling things with their eyes. Learning to operate the device reliably is the
most important thing at first. It's about getting used to controlling things with the
eyes in peace without the pressure of immediately having to hit the 'right' fields.

Controlling a communication aid or computer with the eyes is very taxing and
unfamiliar at first. Short units and breaks are important at the beginning. It's a good
idea to end a unit at a point where usage is working well. It's important to be
patient, as it takes some time to learn how to operate gaze control. All those
involved must learn how to operate it. It's therefore helpful to share progress with
each other. As well as communicating through gaze control, it's good to bear in
mind that there are other forms of communication, such as body language and
meaningful looks or other aids.

Tips for advanced work with a communication aid with gaze control—developing security strategies

- Developed or individualised page sets (communication interfaces) should be regularly saved to another data storage device (for example, to a USB stick, an external drive or another computer).
- The communication aid's set-up function is password-protected where possible. Thus, all the user's attendants can be sure they won't delete or change something by mistake.
- A printout of the page sets being used or another non-electronic communication aid (for example, an alphabet board) must always be on hand. If there are technical problems, the person whose communication is being supported will need this alternative.

In gaze control, an individual trigger method is chosen ('dwell mode', blinking or an exterior button). This is usually put in place during familiarisation. It can make sense to adjust these triggers later. Thus, the trigger time might be shortened or lengthened. In any case, it's a good idea to play around a bit with the trigger time. If a new interface is to be explored, a longer trigger time can be supportive. However, an overly long trigger time delays mutual dialogues and is potentially frustrating and tiring. If the user isn't yet familiar with the principle of cause and effect, it's good to have a good trigger time and to evoke a quick reaction to the user's gaze.

3.8 How Does Environment Control Work?

As well as activating a communication aid or computer, a person with severely impaired motor function can use gaze control to operate various domestic devices. Interfaces are offered on the communication aid or computer that, for example, allow light switches, the TV, the CD player, curtains and so on to be operated. These devices are connected to the person's communication aid or computer by infra-red. Using infra-red or Bluetooth, it's also possible to operate a mobile phone and, for example, write text messages. This requires the appropriate interface to be installed.

The interface for controlling a person's surroundings can be adjusted to the requirements and potential of the person's living environment. With environment control, people who find it impossible or very difficult to manage their surroundings because of severe motor function impairment are able (once again) actively to organise their everyday lives. A couple of examples are pictured and explained below (Figs. 3.8, 3.9 and 3.10).

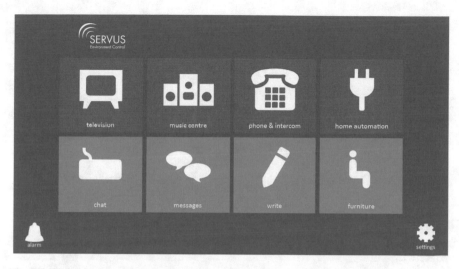

Fig. 3.8 This interface can be used with gaze control on a computer or communication aid. It shows different categories adapted to the living environment of the patient. With one click, various devices can be called up and operated or switches can be activated. *Source* https://grids. sensorysoftware.com/en/sensory-software/environment-control. Sensory Software, Online Grids, Servus, Smartbox

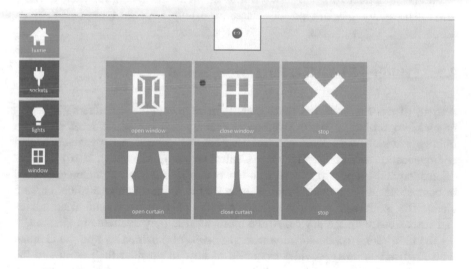

Fig. 3.9 It is possible to open or close windows, doors and curtains or shutters, or to activate certain targeted electrical sockets. *Source* https://grids.sensorysoftware.com/en/sensory-software/ environment-control. Sensory Software, Online Grids, Servus, Smartbox

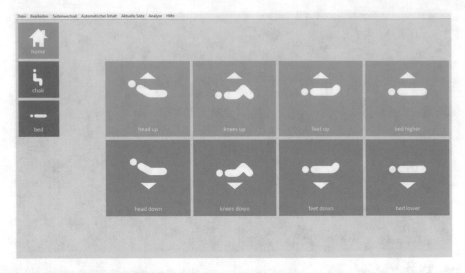

Fig. 3.10 The person also has the option of changing the position of their bed or armchair. *Source* https://grids.sensorysoftware.com/en/sensory-software/environment-control. Sensory Software, Online Grids, Servus, Smartbox

Aid manufacturers provide various forms of environment control. There are now also interfaces for special types of device, for example special types of TV.

3.9   Training Materials for Gaze Control

As part of the EyeTrack4all (gaze control in augmentative and alternative communication) project at Alice Salomon University of Applied Sciences in Berlin training materials were developed that are designed to make using gaze-controlled communication aids easier. The 'Gaze Control Guide' (Kollak et al. 2016) provides orientation and support to anyone who is wondering if using a communication aid or computer with gaze control makes sense for them, for a family member or for a client. The guide also provides assistance with applying for communication aids and firsts steps with the device. At the moment, it is only available in German.

The film '*Better communication with gaze control: positioning tips*' (in German with English subtitles) provides examples of how to position the person and the device. A second part (without subtitles) presents a 'personal budget'. That is a particular funding model for care services or aids to which people in Germany have a legal entitlement.

3.10 Summary

Using gaze control to operate a communication aid or computer provides vital access to self-determination and participation for people who cannot express themselves or move on a temporary or permanent basis because of illness. This is an essential key to better quality of life. Gaze control users may communicate their wishes and needs to those around them and thereby influence their situation, or design their own environment with environment control. The communication aid or computer's various functions and interfaces can be adjusted to meet individual needs. If it is possible for a person to communicate well with gaze control and influence their own situation, this can have a positive effect on rehabilitation and prevention.

This chapter was produced as part of the research project 'EyeTrack4all—enlarging the user group for augmentative and alternative communication using gaze control'. The project was based at Alice Salomon University of Applied Sciences in Berlin, ran for three years and was sponsored by the German Federal Ministry of Education and Research. The project's aim was to improve use of gaze control by researching how the technology is used in everyday life and producing training materials. The project was conducted in cooperation with a company that develops gaze controls (alea technologies GmbH) and a neurological rehabilitation clinic for children and young people (Hegau Jugendwerk). Intensive interchange with experts in the augmentative and alternative communication area rounded the work off. Observing the use of gaze controls in everyday life gave the researchers the opportunity to become familiar with the tricks and difficulties associated with implementing assistive technology (Saborowski et al. 2015). They focused on various issues. What does a communication situation where gaze control is used look like? What strategies have users and communication partners developed to implement gaze control in a positive way? What is the effect on those around the user? These observations were supplemented by knowledge from experts working in the area of augmentative and alternative communication.

Acknowledgments The authors all belonged to the EyeTrack4all project team.

References

Andres P (1996) Die Bedeutung der Positionierung für eine erfolgreiche Unterstützte Kommunikation [Importance of positioning for successful supported communication]. In: Arnusch G, Gesellschaft für Unterstützte Kommunikation (eds) „Edi, mein Assistent" und andere Beiträge zur unterstützten Kommunikation ['My assistant, Edi' and other contributions on augmentative and alternative communication]. Reader der Kölner Fachtagungen. Verlag Selbstbestimmtes Leben, Düsseldorf, pp 290–299

Blackstone SW, Pressman H (2016) Patient Communication in Health Care Settings: new opportunities for augmentative and alternative communication. Augment Altern Commun 32 (1):69–79. doi:10.3109/07434618.2015.1125947

Convention on the rights of persons with disabilities. Available via DIALOG. http://www.un.org/disabilities/convention/conventionfull.shtml. Accessed 26 Aug 2016

Fager S, Bardach L, Russell S et al (2012) Access to augmentative and alternative communication: new technologies and clinical decision-making. J Pediatr Rehabil Med 5(1):53–61. doi:10.3233/PRM-2012-0196

Kollak I, Nuß C, Saborowski M (2016) Handreichung Augensteuerung: Hilfestellung für Vorüberlegungen, Planung und Einsatz einer Augensteuerung in der Unterstützten Kommunikation [Gaze Control Guide: assistance with preliminary considerations, planning and implementation in augmentative and alternative communication]. Available via DOALOG. https://nbn-resolving.org/urn:nbn:de:kobv:b1533-opus-1334. Accessed 18 Aug 2016

Majaranta P, Aoki H, Donegan M et al (eds) (2012) Gaze interaction and applications of eye tracking: advances in assistive technologies. IGI Global, Hershey

Saborowski M, Grans AL, Kollak I (2015) Wenn Blicke die Kommunikation steuern: Beobachtung einer Augensteuerung im Alltag [When gaze controls communication: observations of gaze control in everyday life]. In: Antener G, Blechschmidt A, Ling K (eds) UK wird erwachsen: Initiativen in der Unterstützten Kommunikation [The UK grows up: initiatives in augmentative and alternative communication]. von Loeper, Karlsruhe, pp 370–383

Chapter 4
Caring TV—for Older People with Multimorbidity Living Alone: Positive Feedback from Users in Berlin and Rural Mecklenburg-West Pomerania

Stefan Schmidt and Anke S. Kampmeier

4.1 Background

Demographic change is a challenge for Germany's ageing society. Old age structures will change under the influence of decreasing birth rates and increased life expectancy (Hoffmann et al. 2009). Instances of multimorbidity increase considerably with old age. According to the DEAS (German Survey of Old Age), three quarters of those aged over 65 reported having more than one disease. Just under half of those aged 40–85 (46 %) already have two to four concurrent diseases, and one in five of those over 85 has five or more concurrent diseases. With increasing age the risk of needing care grows, and this can bring not just a loss of health but also diminishing independence (Wurm et al. 2010).

The population of Germany is 81.8 million. Currently around 2.6 million of those people need care, of which the great majority (around 80 %) is aged 65 and over. Of those aged over 90, about one half (59 %) need care. The number of women needing care increases markedly after the age of 80 compared to men of the same age. Just under 70 % (1.86 million) of those needing care are looked after at home. Of them, 1.3 million are cared for by family members. Six hundred and sixteen thousand of those in need of care receive support from home care services (Federal Statistical Office 2015). Thus the majority of care recipients live in their home. However, it is evident that the number of single-person households also increases with age (Menning 2007, 2008).

Loneliness and social isolation also increase with age, as social networks change, and partners, relatives and friends are no longer there. Social networks are

S. Schmidt (✉) · A.S. Kampmeier
University of Applied Sciences Neubrandenburg, Neubrandenburg, Germany
e-mail: stefan.schmidt@ash-berlin.eu

© The Author(s) 2017
I. Kollak (ed.), *Safe at Home with Assistive Technology*,
DOI 10.1007/978-3-319-42890-1_4

smaller in old age (Boeger et al. 2016; Petrich 2011), and so older people tend to live alone, despite multimorbidity and a growing need for support and care.

The question therefore arises of how to reach out to older people with multi-morbidity who live alone and how they can obtain the information they need. Increasingly, technical aids are used to reach people living at home. One of the main issues here is how to ensure that older people participate in the community despite medical conditions or precisely because of them.

4.2 Selected Research Results on the Use of Technology

The use of technological tools for home care can be divided into three categories: (1) improving surroundings and making them safer, (2) controlling the provision of medical care, (3) using tablets, text messaging and other suitable transmission formats.

1. There is already a number of studies available on Ambient Assisted Living Systems for people who are chronically ill. These studies aimed to improve the safety of older people in their living environment and increase their autonomy and self-sufficiency by using age-appropriate support systems. Studies show that technical support has a positive effect on fall prevention (Vaziri et al. 2016), that the risk of falling can be reduced (Gschwind et al. 2015), that exercise and training programmes can combat frailty (Mehrabian et al. 2015) and that technology can lead to an overall improvement in safety in the home (Pearce et al. 2012).
2. In further studies, support was given to chronically ill older people by measuring their vital signs. This information was transmitted and evaluated by tele-medicine. The projects aimed to react to any health problems arising in a timely and appropriate manner. The underlying motivation was to extend the length of time for which older people were able to live independently and safely in their usual surroundings (Falcó et al. 2013; Jara et al. 2011; Norgall and Wichert 2013; Spitalewsky et al. 2013). Other studies were conducted into facilitating self-monitoring of blood sugar levels and using software tools to transmit blood sugar data to practitioners (Stegmaier et al. 2013). Technical tools play an important role in supporting chronic medical conditions such as diabetes (von Bargen et al. 2013). Counselling programmes available by telephone can help to build trust among users and motivate those with chronic conditions to change their everyday behaviour.
3. Activation using tablet PCs has positive results with regard to communication, interaction and well-being for people with dementia who live in care institutions (Nordheim et al. 2014). In Finland, a video conference programme featuring interaction between people needing support and care and healthcare professionals was used. Acceptance by users is high (Raij and Lehto 2008; Lehto 2013). A further study showed that using tablet PCs can improve quality of life for the elderly by increasing contacts with family and friends (Damant et al. 2016).

4.3 Our Study

Our study analyses a four-pillar model that offers the following services: socialisation programmes for the elderly (i.e. affordable activities such as keep fit and
games afternoons); supervised living (visiting services, help with housework,
accompanied walks); and professional care advice provided by nurses and social
workers from care support points. We have analysed these programmes in a study
and combined them with a new tool called Caring TV (see Fig. 4.1).

Caring TV helps the elderly to access advice from professional care personnel
and social workers through a video conference programme, to engage with their
peers in live chats and to download relevant material on old age such as funding for
care and fall prevention options through a Dropbox folder. Caring TV was facili-
tated by using an iPad which was supplied to study participants.

Study facts:

- The study was an intervention study carried out in two different states in
  the Federal Republic of Germany: one a large city, in this case Berlin, and
  the other a rural area, in this case the state of Mecklenburg West
  Pomerania.
- The study covers the period from October 2011 to December 2014.
- The study was sponsored by the Federal Ministry of Education and
  Research.
- The Ethics Committee has given it a positive vote.

Project partners:

- Diakoniewerk Kloster Dobbertin gGmbH Haus Auf dem Lindenberg,
  Neubrandenburg, Germany (nursing home)
- Miteinander Wohnen e.V., Berlin, Germany (an association that provides
  socialisation programmes and supervised living)
- Neuwoba, Neubrandenburger Wohnungsbaugenossenschaft eG,
  Neubrandenburg, Germany (housing association)
- Paritätischer Wohlfahrtsverband Berlin e.V., Berlin, Germany (welfare
  association)
- Pflegestützpunkt Lichtenberg, Berlin, Germany (care support point)
- Sana Klinikum Lichtenberg, Berlin, Germany (hospital)
- Videra, Oulu and Helsinki, Finland (technology provider)

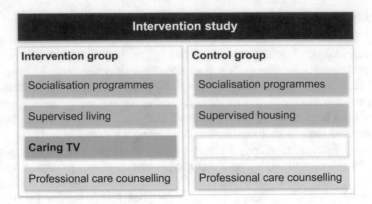

Fig. 4.1 Four-pillar model of elderly

4.3.1 Group Discussions on Specific Topics for Caring TV

Prior to the launch of Caring TV, we held five group discussion in 2012 with a total of 77 people aged between 65 and 90. Fifty-one of these were female and 26 male. We wanted to find out what older people were interested in and what they expected from live chats of this kind. We recorded and subsequently coded the group discussions (Saldana 2013).

Analysis of the group discussions showed the participants were interested in communication. This is not surprising since we visualised the Caring TV programme as enabling older people to communicate with each other from the outset. There was a lot of interest in talking to peers and exchanging views on subjects such as old age and coping with life events such as the loss of a partner. At the same time, it was also important to enable exchange with professionals. Here the emphasis was on issues such as funding for care, care aids, funeral arrangements, illnesses and their progression and concerns such as sleeping problems. Also of interest were leisure activities, with games, quizzes and cooking programmes being very popular. Some participants expressed an interest in using video conferences such as Caring TV for doing exercises, subject to the exercises being easy and suitable for doing at home.

Feedback from the group discussions led to three types of Caring TV programmes being made available: (1) programmes about activities, (2) programmes that supply information and (3) programmes that facilitate communication (see Fig. 4.2).

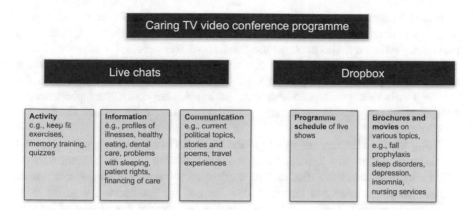

Fig. 4.2 Caring TV video conference programme

4.3.2 *Acceptance of Tablet PCs*

Prior to the Caring TV intervention, we examined tablet PCs. We tested different kinds of tablet PC with older people before deciding on the iPad. The aspects we looked at were handling, optical presentation, simplicity of use and options for programme use. A total of 13 people were observed, of whom eight were women and five were men. They were aged between 53 and 83. Ten of them said that they were very interested in technology. At the time of the trial, six had a computer of their own and used the Internet regularly. Ten had a mobile phone.

The weight of the iPad was rated very positively by five participants. Eleven participants found it easy to hold in the hand. An iPad holder was used successfully by nine participants.

Thirteen of the participants rated the size of the symbols on the iPad as good. Four had problems with the colour settings. For this reason, we later adjusted the contrast settings. Ease of identifying the symbols was rated good by nine people. However, three people found some symbols impossible to identify, and one person found a lot of the symbols impossible to identify. Finding symbols and apps was a little more difficult. Six people could open apps after 'searching around a bit' while four others had difficulty opening apps. Three were only able to find and open the apps with the help of the observer.

No one had difficulty with charging the battery. Nine people found the read aloud option on the iPad very useful. Volume and quality of tone were rated good. The biggest problem we observed was that the default setting on the iPad and apps was English. Some people found the iPad charge cable much too short. Others had trouble with the touch function but got accustomed to this after a while.

The group discussions about Caring TV programmes and the observations we made about the use of tablet PCs were important preparatory work for our study.

Based on these themes, programme concepts for Caring TV were developed. Back-up material was produced for the programmes. This included videos on preventing falls, an audio book on the funding of services, and a number of brochures on sleeping problems in old age and healthy diet. Based on the themes, a number of appointments was arranged with experts who would then appear on Caring TV. In this way, physiotherapists, dieticians, care counsellors, pharmacists, GPs, dentists and insurance experts were recruited to take part in Caring TV programmes.

Using what we learnt when testing tablet PCs with older people, we developed a training concept on how to handle the iPad. Our training sessions concentrated on explaining how to use the iPad and on explaining the apps for the 'Vidyo' video conference programme and Dropbox. We backed up the training sessions with appropriate material. This consisted of a two-page pictorial description of how to operate the iPad and how the apps work.

4.3.3 Study Aim and Questions that Arose

The study asked questions about how people over the age of 65 use these services, what influence the services have on their quality of life and their ability to remain independent and whether the services make it possible for them to take part in life of the community.

We know from previous studies that people prefer to remain living in their own homes for as long as possible (BMG 2011). We also know that they are often not well informed about care programmes and unaware of what is available to them. It is also known that preventive programmes and counselling can reduce the length of time spent in hospitals and care homes, help avoid falls and have a positive influence on mortality (Leung et al. 2010; Parsons et al. 2012). Such programmes can improve quality of life and give people prone to depression the feeling that they are getting support (Gensichen et al. 2009, 2011). But we also know that the number of single households increases as people grow older, the risk of becoming dependent grows, older people are increasingly lonely and relationships with other people grow fewer as people age (Boeger et al. 2016). The use of technological tools can have a positive effect on older people's lives. That is what our study was all about.

In this paper, we mainly look at the users' level of experience with technology and how users of Caring TV reacted to it. The paper concentrates mainly on Caring TV as an intervention.

Publications about informal care programmes, assisted living and professional care counselling have been produced (Schmidt and Kraehmer 2016; Schmidt and Luderer 2014). Further publications evaluating data about quality of life are planned.

The central issues focussed on were:

- What information is there about older people and technology?
- What was the response to the moderated video conference facility and how useful is it for older people with multimorbidity who live alone?
- How did users of the moderated video conferencing facility accept the programme?

Evaluation of the data and this contribution to *Safe at Home* were completed as part of the lead author's dissertation.

4.3.4   Methods

4.3.4.1   Sample Recruitment

We placed advertisements in newspapers to attract people to take part in our study. Partners cooperating in our study also helped us to gain access to older people. The criteria for participation were that people should be living alone at home, be 65 or over and already be using services provided for the elderly. They had to be prepared to take part in the study, and their written consent was, of course, required.

We started a newspaper campaign in January 2013 using daily papers in Berlin and Mecklenburg West Pomerania. We aimed to find 40 participants, 20 from Berlin and 20 from Mecklenburg West Pomerania. We divided these into two groups using randomisation: an intervention group and a control group. The 20 people in the control group were already using socialisation programmes for older people, supervised living or professional care advice. The 20 people in the intervention group also used Caring TV. We gave them an iPad with a video conference programme so that they could take part in our live shows. We instructed each participant on how to use the iPad. The training sessions were very intensive and took up a lot of time. Our sample group totalled 40 people, 33 women and seven men aged between 65 and 92 (see Table 4.1).

- Criteria for participation in our study
- People living alone at home
- Aged 65 or over and already using services provided for the elderly
- Written consent to take part in the study

Table 4.1 Characteristics of the sample

		Intervention group (n = 20)	Control group (n = 20)
Gender	Female (%)	16 (80)	17 (85)
	Male (%)	4 (20)	3 (15)
Age range (M)		66–92 (77)	65–89 (75)
Contact via	Newspaper	11	11
	Cooperation partner	9	9
Care level	0 (care need < 90 min/d)	14	11
	I (care need 90–180 min/d)	4	8
	II (care need 180–300 min/d)	2	1
	III (care need > 300 min/d)	0	0
Status	Single	3	0
	Widowed	15	14
	Divorced	1	6
	Married	1	0
Barthel index (M)	Baseline	55–100 (91)	55–100 (91)
	t1	45–100 (92)	80–100 (92)

4.3.5 Data Collection

In the team, we developed an interview guide. The interview guide covered aspects of everyday life, social contacts, need for support and experience with technology and technicians. It was pre-tested for comprehensibility and usability. Suggestions derived from the pre-tests were discussed by the study team and incorporated once a consensus had been reached.

We used the Barthel Index to estimate independence. The Barthel Index is easy to use and simple to apply and gives a good overview of people's ability to care for themselves. It has been tested for validity and reliability in a number of studies (Mahoney and Barthel 1965; Luebke et al. 2004).

In addition, we used the FLQM questionnaire (Fragebogen zur Lebensqualität multimorbider älterer Menschen; Questionnaire on quality of life among older people with multimorbidity) for assessing subjective quality of life. The participants were asked to name areas in which quality of life played a role. Then we asked them how satisfied they were with this area of their life. Our aim was to obtain a figure: 1 no further improvement was necessary, 2 equalled very satisfied, 3 satisfied, 4 rather dissatisfied, 5 very dissatisfied and 6 things couldn't get any worse. We then asked participants how important this area was to them, again obtaining figures ranging from (1) very important to (6) relatively unimportant (Holzhausen et al. 2010; Holzhausen and Martus 2012) (Fig. 4.3).

We also called the participants in the intervention group every 3 weeks to ask what they were using their iPad for, what they thought of Caring TV, what we could

Life domains	Satisfaction						Importance					
My health (high blood pressure)	1	2	3	4	5	6	1	2	3	4	5	6
My husband (unemployed)	1	2	3	4	5	6	1	2	3	4	5	6
My daughter (lives away)	1	2	3	4	5	6	1	2	3	4	5	6
My work (home office days ?)	1	2	3	4	5	6	1	2	3	4	5	6
My appearance (hair loss)	1	2	3	4	5	6	1	2	3	4	5	6

Fig. 4.3 Example FLQM questionnaire (modelled after Jaccarini et al. 2015, p. 56)

Fig. 4.4 Overview: data collection

Randomization		
	Intervention group (n=20)	Control group (n=20)
March – May 2013	Baseline (interviews, FLQM, Barthel index)	Baseline (interviews, FLQM, Barthel index)
April – December 2013	Intervention Caring TV (phone calls every three weeks)	
October – December 2013	t1 (interviews, FLQM, Barthel index)	t1 (interviews, FLQM, Barthel index)

do better and what the problems were. Over the entire period we made 10 telephone surveys.

Data were collected once at the start of the intervention and again at the end (see Fig. 4.4).

Data collection:

- We used qualitative interviews and quantitative measuring tools. Data on quality of life, satisfaction, independence and self-determination were collected in two phases from people aged over 65 living in their own homes.
- Our instruments for quantitative and qualitative data collection were:
- interview guide (covering aspects such as everyday routines, existing contacts with other people, experience with services already being used and interest in technology)
- Barthel index to measure independence
- FLQM questionnaire for assessing subjective quality of life
- We called the participants in the intervention group every 3 weeks to ask what they were using their iPad for, what they thought of Caring TV, what we could do better and what there were problems.
- Over the entire period we made 10 telephone surveys.
- Data was collected once at the start of the intervention and then again at the end.

4.3.6   Data Analysis

The interviews and telephone surveys were written up in full, and pseudonyms were used for the participants. Analysis was conducted using a several-stage coding process. First the material was openly coded. These codes were then placed into categories. The categories emerged deductively parallel to the guide. Some categories emerged inductively from data material. Code notes were also written to record initial working results. Evaluation of the interviews was carried out with MAXQDA11 software.

The Barthel Index data and those from the FLQM were mainly counted according to frequency.

The study team held several meetings about the intersubjectivisation and plausibility of the data analysed. The discussion about the results of the analysis led to a communication validation of the results.

4.3.7   Ethics

The Medical University of Greifswald, Germany, endorsed the study.

Participants were given comprehensive written information in advance on the study, its goals and the reason for the research. Participants in the study had a period of about 1 week to read the written information and could clarify any queries with a nominated contact person by telephone.

It was explained to the participants that participation in the study was voluntary, they had the right to withdraw without giving a reason and participation did not entail any risks or disadvantages. Prior to the data being collected, all participants in the study received written confirmation that their data would be kept confidential and pseudonyms would be used. The Caring TV programmes were not recorded. Participants included in the study gave informed consent. Consent was given voluntarily and was obtained in writing by the study team.

4.3.8   The Caring TV Intervention

In the period between April and December 2013 we did 105 caring TV shows. We ran them three times a week, and each broadcast lasted 45 min. Of 105 Caring TV shows, 22 were about activity, 39 shows provided information and 44 focused on communication.

Usually between one and four people took part. In one case, eight people participated.

The participants were sent a programme schedule 4 weeks in advance via Dropbox. We also made brochures on various topics available in Dropbox. We broadcast and presented Caring TV from the university. The shows were anchored by professional care staff and social workers from the study team.

4.3.9   Selected Findings of Our Interviews

4.3.9.1   Previous Experience with the Technology

Some participants had experience with PCs and the Internet prior to our study. They had email and used their computer to communicate with family and friends. None of the participants had any prior experience with tablet PCs or video conference programmes such as Skype, yet they still hoped to get a lot out of participating in video conferences. One participant summed it up by saying they hoped '...*that I might make social contacts, exchange views with other people, visit each other*' (6, 369). Others were interested in receiving information that they would be able to use in their everyday lives. Issues involving care and therapy were often specifically mentioned, but the main focus was on contact with peers. Study participants were eager to try out Caring TV and to use it to get to know other people.

4.3.10   Scheduling of the Shows

Most participants founds the scheduling of the shows good, one reason being that they could decide whether to take part at 10 am or 2 pm. Some felt an obligation to take part in Caring TV and felt guilty if they could not do so because of other commitments. This came out in statements such as: '... *three times a week was okay. You just had to get used to the idea of having one more date to keep*' (14, 221–222).

It emerged from our interviews that participants did not want to forgo contact with others and were also concerned about missing programmes that imparted information. The fact that we regularly repeated shows dealing with specific themes (i.e. funding of care, care support tools, healthy diet, oral hygiene, and activities such as memory training and exercise) was viewed positively. There was frequent feedback on using Caring TV to engage with other people. Two people commented that they would have liked Caring TV shows at weekends and in the evening. All participants were happy that the Caring TV shows were 45 min long.

4.3.11   Usefulness in Everyday Life

The usefulness of the information provided on the programmes and in the back-up materials available to download from Dropbox was generally assessed as very high. Depending on their individual interests, participants found exchange with experts such as care counsellors, physiotherapists, dentists and pharmacists helpful. This was largely due to the fact that Caring TV supplied general information on the subject (e.g. what to do about sleeping problems) followed by an opportunity to ask the experts questions. This was useful because access was easy and fast. Group discussions also prompted new questions and provided ideas on overcoming problems. The point was users could find answers to their questions by interacting with experts and their peers. For example, one person commented that: '...*by interacting with other people you always learn something you can use in everyday life*' (4, 300).

The users enjoyed programmes that focused on activities. Memory training, games and physical exercise were very popular. After one show, some participants thought about quizzes and games that they would like to do with the others. The shows were designed to be participatory, and active participation was explicitly encouraged. However, only a few users made an active contribution to the way shows were structured.

Some users used their iPad after the show to research some of the themes covered, as shown in the following example: '*Afterwards I sometimes looked something up, living wills, for example[....], and that was very useful to me.*' (3, 352–366).

Other users used their iPad for things other than Caring TV and our Dropbox folder options, for example to take photos, look at photos and read books.

For many participants the iPad became an important part of their daily routine because: '*it's very handy and light, and you can always have it to hand*' (14, 339). An important benefit that emerged in the interviews was that access to experts was easier and contact with peers had been improved.

4.3.12 Problems

The biggest problem was the poor Internet connection. If the connection was poor, it was not possible to take part in the live Caring TV shows. Two participants were particularly affected by this. Their Internet connection was regularly interrupted, and the quality of sound and pictures was poor. Unfortunately, not much could be done to change this during the study because high-speed Internet was not available where the participants lived. For most participants, difficulty with the touch function could be remedied with intensive training. Two participants had no further interest in taking part in programmes because they were of no further use to them (33, 909).

4.3.13 The Future

Many participants could imagine taking part in Caring TV programmes outside our study. Caring TV made it possible for them to contact peers and experts more easily. Some participants commented that the programmes needed a presenter: '*You need someone to organise and coordinate the shows, otherwise it won't work*' (9, 397–398).

Only a minority of those participating were prepared to pay for Caring TV so that it could be continued. Although participants' usage of tablet PCs, video conference software and Caring TV was high, few participants were prepared to pay money for them. Participants instead favoured funding by long-term care insurance and building societies.

4.4 Conclusion

It is encouraging that the services provided by our Caring TV shows were well received and that the participants reported high usage.

In summary, our study went according to plan and was successful. Discussions with project partners and experts from theoretical and practical disciplines showed our project was relevant for securing need-oriented and client-driven care and provision for older people in their own homes. It was also clear that the study had aroused great interest among potential users and professional from the health, care and social work sectors and that its aims were widely accepted. Knowledge gained from the study could potentially be used as a basis for new service programmes for the elderly. Caring TV supports exchange with experts and promotes contact among peers. These results are encouraging despite or rather because of multimorbidity in older people, because they show that loneliness and social isolation can be prevented. Ideally, building societies, assisted living services and home care providers would take responsibility for marketing these services. The outlook for this is very promising, particularly as the partners and participants in the project at Neubrandenburg University saw high market potential.

Acknowledgments We are grateful to all the study participants. In addition, we would like to thank our study team members. This study was funded by the Federal Ministry of Education and Research under Grant Number 17S16X11. The contents of this publication are the sole the responsibility of the authors.

References

BMG—Bundesministerium für Gesundheit (2011) Abschlussbericht zur Studie "Wirkungen des Pflege-Weiterentwicklungsgesetzes". Bundesministerium für Gesundheit, Berlin
Boeger A, Wetzel M, Huxold O (2016) Allein unter vielen oder zusammen ausgeschlossen: Einsamkeit und wahrgenommene soziale Exklusion in der zweiten Lebenshälfte. In: Mahne K, Wolff JK, Simonson J et al (eds) Altern im Wandel: Zwei Jahrzehnte Deutscher Altersurvey (DEAS). Deutsches Zentrum für Altersfragen, Berlin
Damant J, Knapp M, Freddolino P et al (2016) Effects of digital engagement on the quality of life of older people. Health Soc Care Community. doi:10.1111/hsc.12335

Falcó JL, Vaquerizo E, Lain L et al (2013) Am i and deployment considerations in aal services provision for elderly independent living: The monami project. Sensors 13(7):8950–8976. doi:10.3390/s130708950

Federal Statistical Office (2015) Care Statistics 2013. Pflege im Rahmen der Pflegeversicherung. Deutschlandergebnisse. Statistisches Bundesamt, Wiesbaden. Available via DIALOG. https://www.destatis.de/DE/Publikationen/Thematisch/Gesundheit/Pflege/PflegeDeutschlandergebnisse5224001139004.pdf?__blob=publicationFile. Accessed 11 Aug 2016

Gensichen J, von Korff M, Peitz M et al (2009) Case management for depression by health care assistants in small primary care practices: a cluster randomized trial. Ann Intern Med 151(6):369–378

Gensichen J, Petersen JJ, Karroum T et al (2011) Positive impact of a family practice-based depression case management on patient's self-management. Gen Hosp Psychiatry 33(1):23–28. doi:10.1016/j.genhosppsych.2010.11.007

Gschwind YJ, Eichberg S, Ejupi A et al. (2015) ICT-based system to predict and prevent falls (iStoppFalls): results from an international multicenter randomized controlled trial. Eur Rev Aging Phys Act 12. doi:10.1186/s11556-015-0155-6

Hoffmann E, Menning S, Schelhase T (2009) Demografische Perspektiven zum Altern und zum Alter. In: Böhm K, Tesch-Römer C, Ziese T (eds) Gesundheit und Krankheit im Alter. Beiträge zur Gesundheitsberichterstattung des Bundes, Robert Koch Institut, Berlin, pp 21–30

Holzhausen M, Martus P (2012) Validation of a new patient-generated questionnaire for quality of life in an urban sample of elder residents. Qual Life Res 22(1):131–135. doi:10.1007/s11136-012-0115-9#

Holzhausen M, Kuhlmey A, Martus P (2010) Individualized measurement of quality of life in older adults: development and pilot testing of a new tool. Eur J Aging 7(3):201–211. doi:10.1007/s10433-010-0159-z

Jaccarini GA, Kollak I, Schmidt S (eds) (2015) Just in case—care and case management in malta. MMDNA, Malta

Jara AJ, Zamora MA, Skarmetwa AF (2011) An internet of things—based personal device for diabetes therapy management in ambient assisted living (AAL). Pers Ubiquit Comput 15 (4):431–440. doi:10.1007/s00779-010-0353-1

Lehto P (2013) Interactive caring TV supporting elderly living at home. Australas Med J 6(8):425–429. doi:10.4066/AMJ.2013.1800

Leung AY, Lou VWQ, Chan KS et al (2010) Care management service and falls prevention: a case-control study in a chinese population. J Aging Health 22(3):348–361. doi:10.1177/0898264309358764

Luebke N, Meinck M, von Renteln-Kruse W (2004) The barthel index in geriatrics. a context analysis for the hamburg classification manual. Z Gerontol Geriatr 37(4):316–326

Mahoney FI, Barthel DW (1965) Functional evaluation: The barthel index. Md State Med J 14:61–65

Mehrabian S, Extra J, Wu YH et al (2015) The perceptions of cognitively impaired patients and their caregivers of a home telecare system. Med Devices (Auckl) 8:21–29

Menning S (2007) Haushalte, familiale Lebensformen und Wohnsituation älterer Menschen. GeroStat Report Altersdaten 02/2007. Deutsches Zentrum für Altersfragen, Berlin

Menning S (2008) Ältere Menschen in einer alternden Welt—Globale Aspekte der demografischen Alterung. GeroStat Report Altersdaten 01/2008. Deutsches Zentrum für Altersfragen, Berlin

Nordheim J, Hamm S, Kuhlmey A et al (2014) Tablet-PC und ihr nutzen für demenzerkrankte heimbewohner. ergebnisse einer qualitativen pilotstudie. Z Gerontol Geriatr 48(6):543–549. doi:10.1007/s00391-014-0832-5

Norgall T, Wichert R (2013) Personalized use of ICT–from telemonitoring to ambient assisted living. Stud Health Technol Inform 187:145–151

Parsons M, Kerse N, Chen M et al (2012) Should care managers for older adults be located in primary care? A randomized controlled trial. J Am Geriatr Soc 60(1):86–92. doi:10.1111/j. 1532-5415.2011.03763.x

Pearce AJ, Adair B, Miller K et al (2012) Robotics to enable older adults to remain living at home. J of Aging Res. doi:10.1155/2012/538169

Petrich D (2011) Einsamkeit im Alter. Notwendigkeit und (ungenutzte) Möglichkeiten Sozialer Arbeit mit allein lebenden alten Menschen in unserer Gesellschaft. Jenaer Schriften zur Sozialwissenschaft Band Nr. 6. Fachhochschule Jena, Jena

Raij K, Lehto P (2008) Caring TV as a Service with and for Elderly People. In: Tsihrintzis GA, Virvou M, Howlett RJ et al (eds) New directions in intelligent interactive multimedia. Springer, Berlin, Heidelberg, pp 481–488

Saldana J (2013) The coding manual for qualitativ researchers. Sage Publications, Los Angeles et al

Schmidt S, Kraehmer S (2016) Care support points of mecklenburg-west pomerania. Results of a scientific analysis. Int J of Health Prof. Available via DIALOG. http://www.degruyter.com/ view/j/ijhp.ahead-of-print/ijhp-2016-0019/ijhp-2016-0019.xml. Accessed 23 Aug 2016

Schmidt S, Luderer C (2014) The work of care support points as experienced by the users—A hermeneutic, interpretative study. Pflegewissenschaft 16(11):631–638. doi:10.3936/1274

Spitalewsky K, Rochon J, Ganzinger M et al (2013) Potential and requirements of IT for ambient assisted living technologies. Results of a delphi study. Methods Inf Med 52(3):231–238. doi:10.3414/ME12-01-0021

Stegmaier P, Harms F, Gaenshirt D (2013) Von der versorgung zur befähigung zum selbstmanagement. Monitor Versorgungsforschung 6(3):23–28

Vaziri DD, Aal K, Ogonowski C et al (2016) Exploring user experience and technology acceptance for a fall prevention system: results from a randomized clinical trial and a living lab. Eur Rev Aging Phys Act 13:6. doi:10.1186/s11556-016-0165-z

von Bargen T, Schwartze J, Haux R (2013) Disease patterns addressed by mobile health-enabling technologies— a literature review. Stud Health Technol Inform 190:141–143

Wurm S, Schöllgen I, Tesch-Römer C (2010) Gesundheit. In: Motel-Klingebiel A, Wurm S, Tesch-Römer C (eds) Altern im Wandel. Befunde des Deutschen Alterssurveys (DEAS). Kohlhammer, Stuttgart, pp 90–117

Chapter 5
Arm Rehabilitation at Home for People with Stroke: Staying Safe: Encouraging Results from the Co-designed LifeCIT Programme

Ann-Marie Hughes, Claire Meagher and Jane Burridge

Keyword Stroke · CIMT · Rehabilitation · Upper Extremity

5.1 Rationale

The worldwide prevalence of stroke in 2010 was 33 million, with 16.9 million people having a first stroke, of which 795,000 were American and 1.1 million European (Mozaffarian et al. 2016). In the UK, the annual incidence of stroke is 150,000, and it represents the second highest cause of disability. Although the provision of rehabilitation therapies is often not the first priority, there are benefits to starting rehabilitation as soon as the patient is ready and can tolerate it (Miller et al. 2010). UK stroke guidelines currently state that patients should receive at least 45 minutes of each appropriate therapy every day at a frequency that enables them to meet their rehabilitation goals (Intercollegiate Stroke Working Party 2016). However, audits report that many people do not receive this, and people with stroke are often prevented from practising tasks themselves as they often have little movement.

Despite receiving existing therapy, only 20–56 % of all stroke survivors regain useful upper limb function after 3 months (Nakayama et al. 1994). That means that annually some 5.6 million people worldwide (265,000 in the US and 300,000 in the EU) fail to regain arm and hand dexterity with resultant increases in personal and societal care costs. In the UK alone, each year 54,000 people are left with hand/arm disability after stroke adding to the 300,000 existing stroke survivors with persistent and disabling upper limb motor impairments. Similar numbers exist in the EU, where over 6 million people with stroke require care and two thirds of these have impairment of their affected arm 4 years post-stroke, resulting in an annual cost of €38 billion (Nichols et al. 2012). This particularly affects the 75 % of younger individuals who want to return to work.

A.-M. Hughes (✉) · C. Meagher · J. Burridge
University of Southampton, Southampton, UK
e-mail: A.Hughes@soton.ac.uk

© The Author(s) 2017
I. Kollak (ed.), *Safe at Home with Assistive Technology*,
DOI 10.1007/978-3-319-42890-1_5

As stroke is an age-related disease, these numbers will increase with an ageing population; the number of people above the age of 65 is expected to increase by two thirds by 2040. Added to this, advancements in acute care, mean that people may be surviving with more serious impairments than before. It is important to consider that the burden of stroke now disproportionately affects individuals living in resource-poor countries. From 2000 to 2008, the overall stroke incidence rates in low to middle income countries exceeded that of high-income countries by 20 % (Truelsen et al. 2007). Due to demographic changes, urbanisation and increased exposure to major stroke risk factors, the prediction is that by 2025, four out of five stroke events will be in people living in these regions (Murray and Lopez 1997).

5.2   Rehabilitation Mechanisms Promoting Recovery

The mechanisms for recovery are poorly understood, but learnt non-use has been identified as an important factor for people with stroke. Simply put, people with stroke often use their unaffected arm to do motor tasks such as reaching out for objects because it is easier and quicker. This means that the affected arm does not get vital movement practice.

Following a stroke, a motor task may be performed in exactly the same way as before, which is termed 'recovery'. This occurs through neuroplastic changes resulting from neuronal activity or connectivity changes. Functional electrical stimulation (FES) is one example of a therapy that works through direct neuro-plastic changes. When the task is performed in a different way, this is termed 'compensation'. This occurs through abnormal muscle or movement synergies that may or may not enhance function or lead to behavioural changes, which may also promote neuroplasticity. Constraint Induced Therapy (CIT) also known as Constraint Induced Movement Therapy (CIMT) (Taub et al. 1993, 1998), in which constraint of the unaffected arm and hand is coupled with intensive training of the hemiplegic limb, is one example of a therapy that aims to promote behavioural changes by encouraging greater use of the hemiplegic limb, as well as neuroplastic cortical changes resulting in return of function (Taub et al. 1993). Simply put, learnt non-use is reduced by restricting activity in the unaffected arm, encouraging use of the affected arm.

To acquire a skill requires varied, goal-orientated practice (Plautz et al. 2000; Nudo 2003), feedback, intensity of practice, and attention and motivation to maximise engagement and effort as well as to comply with the training programme. Sensory motor experiences are associated with neuroplasticity and recovery (Nudo 2003): associated stimuli increase neural excitability (Ridding et al. 2000), and random and varied practice of tasks improves learning and retention in stroke patients (Hanlon 1996). Repetition and practice are critical to skill acquisition

(Ericsson et al. 1993), however, compelling evidence shows that greatest recovery occurs when doing meaningful tasks aimed at acquiring a practical skill, rather than simple repetition of exercises (Krakauer 2005). There is also evidence to suggest that goal-orientated practice using real objects is more effective than virtual tasks (Wu et al. 2000).

Recovery of function is associated with obtaining cortical changes in the brain. From work conducted by Nudo et al. (1996), we know that intensity, challenge and practice are important to stimulate these changes. So how intense does practice of functional tasks have to be? The average number of upper limb repetitions from therapy has been estimated at 30–60 per session (Lang et al. 2009; Kimberley et al. 2010). From animal studies we know that the number of repetitions a day leading to structural neurological change is 400–600 (Plautz et al. 2000; Nudo et al. 1996; Kleim et al. 1998). Carey et al. (2002) reported that people with impaired grasp-and-release secondary to stroke who performed more than 100 repetitive movements per day (1,200 total) of a finger-tracking exercise demonstrated significant cortical reorganisation and functional improvement compared with control subjects.

5.3 Technology at Home—Design and Implementation

Clearly, existing ways of delivering therapy are not achieving effective results, and financial, personnel and time constraints mean that post stroke therapy will be delivered differently in the future. A review conducted by Teasell et al. (2013) found that for stroke patients, home-based and hospital-based outpatient therapy appeared to be equally effective. It has been established that outcomes are closely related to the intensity of training undertaken, and this creates a role for technology that can be used in the homes of people with stroke to promote convenience, adherence and regularity of training, self-management, long-term continuation and reduced clinical contact time. A technology that people with stroke and carers can become familiar with using in an acute setting and then continue to use in their home has the potential to be effective, but it has to be very well designed. To achieve this, it is important that users' needs are considered at all stages of the development and implementation of any technology.

Good design needs teamwork. To be able to co-develop well-designed technologies, teams need increasingly diverse fields of competence to: understand complex home environments, ensure technology choices are driven by neuroscience, make good design choices in collaboration with users, and build simple-to-use, robust, safe, reliable, motivating technologies that are clinically and cost effective. A recent paper argued that a new discipline is being established, that of rehabilitation technologist. Competences were published for an EU MSc in Rehabilitation Technologies (Hughes et al. 2016) aimed at co-educating healthcare professionals with engineers. Practices, methods, and theoretical bases need to evolve to ensure that the emphasis is on people's needs, rather than the technology. In addition, translation mechanisms need to be addressed. Challenges faced in trying to support post-stroke rehabilitation in the home in a 'Motivating Mobility Project' included the extended 'user' network, and the challenge of autonomy and motivation (Fitzpatrick et al. 2010).

Rehabilitation technology for people at home has to be designed to be suitable for use outside a hospital environment. It has to meet the needs of patients and therapists, but also carers and family as well. People may or may not know what they want a technology to do or to look like, and they may need time to consider what is important to them in their social context. In addition, relevant education and training may need to be considered for these people.

It can be challenging to design appropriate materials; there is often a tension between people with stroke wanting to try new challenges at home but stay safe at the same time. Within this environment there may well be issues of practicality versus individuality. There are a number of questions surrounding the design of devices for home use: what is developed may depend on whether the designers view people with stroke as a generic healthcare problem to be solved or as individuals with diverse interests, families and friends. Is the home viewed as the patient's own environment, full of memories and furniture, or an extension of a clinical space? Is technology designed to meet merely clinical goals or might it also address quality of life needs? Additionally, are clinicians viewed as controllers/monitors of care or as guides and advisers?

Other design considerations are likely to include privacy, interoperability, data integrity, infrastructure and power back up, as well as on-going maintenance requirements. The designed rehabilitation devices will also need to meet relevant safety standards. The International Electrotechnical Commission produces standards for medical electrical equipment with a collateral standard for medical electrical equipment and medical electrical systems used in the home healthcare environment (IEC 60601-1-11—Home Healthcare). Home healthcare environments can include dwelling places where patients live and other places where patients are present, for example cars, buses, trains, boats and planes, wheelchairs and, when walking, the outdoors. They also includes nursing homes. The standards cover basic safety and essential performance. Manufacturers who wish to place medical electrical equipment on the European market must apply CE marking to their devices to indicate compliance with applicable European Medical Device directives. In the future, situational awareness of rehabilitation devices may also need to be considered.

Following on from a well-designed device that meets all the required standards, there will be implementation and translational challenges. Manufacturers and researchers need to consider how the device is embedded into everyday life (what is acceptable, useful, engaging and delivers some quality of experience). There may also be process challenges: implementation of new ways of thinking, acting and organising in home-based rehabilitation, such as negotiations with transition services when patients move from hospital to home. There will also be structural challenges around the integration of new systems of practice into existing home settings.

5.4 Perceptions of Existing and Future Arm Rehabilitation Devices

A survey of 123 people with stroke and carers and 296 healthcare professionals from the UK (Hughes et al. 2014) investigated people's perceptions about assistive technologies for arm rehabilitation post-stroke. The survey discovered that in both groups many reported that they did not use or prescribe assistive technologies (56 % of people with stroke, 41 % of healthcare professionals) with a dearth of information, lack of education and difficulty in access being cited as the main reasons.

Where the technologies were used, patients and carers and healthcare professionals were asked to report which technologies they had used or prescribed. Patients and carers and healthcare professionals reported using/prescribing functional electrical stimulation (FES), virtual reality (VR), CIMT and dynamic splints. They were then asked to report which technologies they used or prescribed most frequently. The results are shown in Figs. 5.1 and 5.2.

When asked to respond to the question: 'Thinking about the assistive technology you use most often, do you think this device is safe?', for VR 50 % of people with stroke and their carers responded yes, for dynamic splints 73 %, for CIMT 75 %

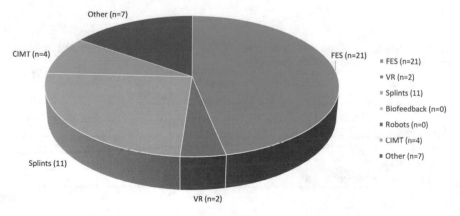

Fig. 5.1 Technologies that patient and carers report using most frequently

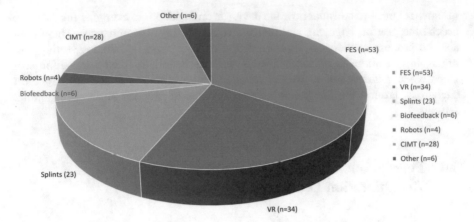

Fig. 5.2 Technologies that healthcare professionals report prescribing most frequently

and for FES 76 %. Healthcare professionals had slightly more confidence in VR (88 %), dynamic splints (74 %), biofeedback (100 %), robots (100 %), CIMT (93 %) and FES (98 %).

The survey also investigated what people believed was important in devices yet to be developed. Table 5.1 shows the top five of the 10 ranked factors. It can be seen that each group gave safety the same priority.

Following on from these survey results, we aim to develop technologies that meet these criteria and bring affordable, effective physical therapy into the homes of people with stroke (ideally with the potential to bridge the transition from hospital to home), allowing them to practise challenging, goal-orientated functional activities with minimal input from a carer or therapist. In this way, these technologies can increase the intensity of rehabilitation without a corresponding increase in clinical contact time. This has the potential to lead to better outcomes, such as reduced impairment, greater restoration of function, improved quality of life and increased

Table 5.1 Rankings of top five factors identified by both patients and carers and healthcare professionals for the design of an ideal AT

Ranking factors	Patients and carers rankings	Healthcare professionals rankings
Easy to set up and use	1	2
Comfortable	2	4
Low risk of harm (safety)	3	3
Durable and reliable	4	5
Good research evidence	5	1

social activity. This in turn will translate into greater independence leading to less dependence on carers and the possibility of returning to work for some.

5.5 LifeCIT: An Example of an Upper Limb Rehabilitation Technology—Research Evidence for CIMT and Clinical Use

As previously discussed, in CIMT learnt non-use is reduced by restricting activity in the unaffected arm, encouraging use of the affected arm. CIMT has been shown to improve arm activity, participation and quality of life in people with a baseline ability to control wrist and finger extension compared with usual care (Taub et al. 1993, 1998), although it is currently unclear if CIMT has any advantage over dose-matched conventional upper limb therapy (Dromerick et al. 2000; Boake et al. 2007). CIMT can be delivered in different forms. The original form was for 3–6 h a day, 5 days a week for 2 weeks, while a modified version was for 1 h a day, 3 days a week for 10 weeks. The modified CIMT intervention appears to result in improvements that are comparable to the original version, although it has not been as extensively tested.

Recent US guidelines (Winstein et al. 2016) recommend task specific training and CIMT given the high level of evidence for these. Note that the paper makes the distinction between basic Activities of Daily Living (ADLs) 'associated with personal self-care and fundamental mobility' and ADLs involving 'more complex domestic, community, and leisure activities' which it terms IADLs. The recommendations and evidence levels are: Functional tasks i.e. appropriately challenging task-specific training should be practised and frequently progressed (class 1, level of evidence A); ADL training should be tailored to individual needs and the eventual discharge setting (class 1, level of evidence A); IADL training should be tailored to individual needs and eventual discharge setting (class 1, level of evidence B); and CIMT/modified CIMT is reasonable to consider for eligible stroke survivors (class IIa, level of evidence A).

Despite the stated evidence for effectiveness including that from a large RCT with 222 participants, which showed improvement in function maintained at 12 months post-treatment (Wolf et al. 2006) and the recommendations from associated guidelines, CIMT has not translated into clinical practice, due partly to the cost of intensive (minimum 4 h a day for 10 days) one-to-one treatment. Healthcare professionals consider the resources required to be a prohibitive barrier (Hughes et al. 2014). A second barrier to adoption is concern about the comfort and safety when the unaffected limb is constrained, especially in the home environment (Hughes et al. 2014).

Fig. 5.3 C-MIT mesh *upper side* for *right* hand

Fig. 5.4 Safety demonstration using C-MIT to hold a walking stick

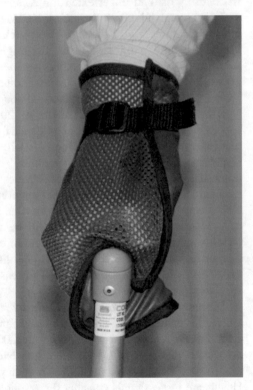

5.5.1   Addressing Translation Factors: Development of Glove—Safety, Comfort and Evidence

An affordable, comfortable hand constraint mitten (C-MIT) that allowed gross movement such as restricted manual tasks was developed (see Figs. 5.3 and 5.4). The design of the C-MIT, enables the unaffected arm to maintain its role in balance and bilateral tasks. The high friction palm enables people to effectively grip a walking aid, ensuring safe mobility. Inside the mitten, a plastic insert and internal padding restricts thumb opposition and wrist activity, preventing too many tasks being undertaken. The mesh upper-side and internal structure increases air circulation to prevent heat build-up, improving comfort.

A research project examined whether clinical benefit would be gained if the C-MIT were used in conjunction with a 2-week, home-based exercise programme without intensive one-to-one therapy. Assessments were carried out prior to the intervention, at completion of the intervention, and repeated 2 weeks later. Improvement was found in upper limb function during the intervention phase after correcting for background recovery, but in their feedback patients reported a lack of motivation to wear the C-MIT and to comply with the exercise programme (Burns et al. 2007). Clearly motivation also needed to be addressed to support the home use of this CIMT.

5.5.1.1 LifeGuide—Motivational Software

Interventions designed to influence people's behaviour ('behavioural interventions') are a fundamental part of daily life, whether in the form of personal advice, support and skills training from professionals (e.g. educators and doctors) or general information disseminated through the media. Personal advice and support are costly, and information provided through the media is not personalised and does not provide support to help people change their behaviour. Interactive technology allows advice to be 'tailored' and responsive to the needs and preferences of each individual. Intensive on-demand support can be provided for behaviour change in the form of guidance, reminders, personalised feedback regarding progress and overcoming obstacles, help with planning, and opportunities for communication with professionals, family and peers.

In our study we used 'LifeGuide' (https://www.lifeguideonline.org), a set of software resources that allows researchers and clinicians to create and modify programmes of personalised online support for therapy, without the need for costly dedicated programming. The software provides information, advice, education and assistance in making decisions, personalised to the situation, interests and capabilities of the user. LifeGuide employs a variety of techniques to support and sustain behaviour change, such as aids to goal-setting, planning and self-monitoring, skill and confidence-building, cues and reminders (e.g. automated emails and texts), and systems of incentive and social support (e.g. encouraging messages from therapists and family members based on the user's activities). The system supports evaluation of interventions, including online assessment.

It is not about giving instructions with the therapist in charge saying: 'this is what you need to do', but about enabling the patient to take the lead with support and guidance asking: 'what do you want to be able to do?'. The idea is to encourage self-efficacy and independence rather than compliance and dependence.

5.5.2   Co-design of LifeCIT with Patients, Carers and Therapists in a Clinical/Home Environment

To identify and implement the necessary components of this web-based intervention, qualitative 'think aloud studies' were carried out with a purposive sample of four chronic stroke patients. Data collection and analysis were concurrent with intervention development, allowing immediate modification and retesting of intervention components as potential improvements were identified. This was followed by the recruitment of 13 acute stroke patients and their carers/therapists in early 2012 to take part in further 'think aloud' studies.

The results from these studies led to the development of the final version of the website, which provides personalised advice on request and has an embedded assessment based on the Motor Activity Log (MAL) that locates the user in the rehabilitation landscape (see Fig. 5.5).

Users are invited to set targets and goals and establish a programme of mitt wear, exercises and tasks each day. Visualisation exercises and specially designed computer games are available, and feedback on progress is provided. If progress is not as hoped, the programme provides advice on how to adapt behaviour. It includes links to therapists and supporters.

Fig. 5.5   Screen shot which shows the user how they are progressing on the 'journey of recovery'

5.5.3 Research Trial: LifeCIT in the Home

The objectives of the study were to test the feasibility and potential clinical benefit of LifeCIT with stroke patients in their own homes as well as to estimate the variability of outcome measures for a sample size calculation for a phase III trial (these trials compare new treatments with the best currently available treatment). The trial design was based on the EXCITE study (Wolf et al. 2006), using the same primary outcome measure. However, replacing one-to-one therapy with web-based support allowed the duration of intervention to be extended to 3 weeks without additional cost.

For the LifeCIT programme, UK ethical approval was gained from South West NHS Ethics Committee prior to the study commencing. The LifeCIT web-based programme was co-designed with patients, carers and healthcare professionals to be accessible, appropriate for each individual and motivating. Personalised LifeCIT data was secure, as it could be accessed via the users' unique username and password, but as stated previously, the users could nominate 'supporters', such as their healthcare professional, friends or family, to have access to their data and give secure motivational feedback via web-based messages. The programme incorporated a self-assessment of ability, based on the Motor Activity Log (MAL), which users completed when they first logged on to the programme. The 10 tasks for self-assessment were chosen to encompass a range of upper limb movements and were informed by published data from the development of the MAL and Rasch analysis (Uswatte et al. 2006). Personalised functional activities were selected from a menu appropriate to each individual, derived from the MAL assessment. Users could also choose functional activities not included in the menu.

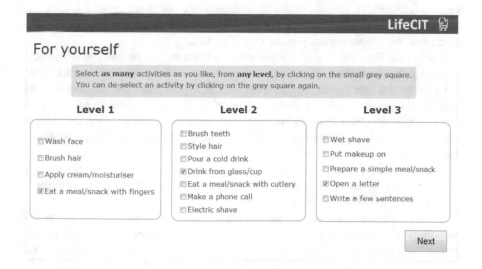

Fig. 5.6 Screen shot of daily activity choices

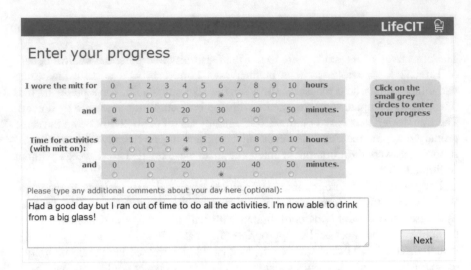

Fig. 5.7 Screen shot of daily progress

Functional activities were classified into 'personal', 'for fun' and 'for the home'. Moving images in GIF format were used to demonstrate the correct way to perform repetitive exercises. Simple written instructions on how to perform the exercises and potential pitfalls were also given. Simple web-based computer games, which could be played using arm movement with the computer mouse, were designed or adapted from commercially available games.

Each week users were prompted to set goals for the time they would wear the C-MIT and perform web-based and functional activities. Users were then asked to log into the LifeCIT website at the start of each day and were given the option of selecting a range of web-based, repetitive exercises and functional activities, before continuing with their set daily goals (see Fig. 5.6).

The system kept a record of the goals the user had set for the week and reminded them what they were each day. They were prompted to amend their selection if they had selected too many or too few to meet their time goal and were given motivating feedback in the form of suggested actions if they were failing to achieve their personal targets. Once they had designed an appropriate programme for the day they were given a choice between printing it, texting it to a mobile phone or leaving it on the computer.

Later each day, users were asked to login again (see Fig. 5.7) to report on adherence to their chosen activities, any difficulties encountered and to complete a second set of repetitive exercises.

These data were stored on the website and could be accessed by their nominated supporters and health care professionals; they could therefore be used to give progress-relevant advice and feedback and to send motivational messages and suggestions for overcoming barriers. Data were also used to monitor use of the system, i.e. patient-reported adherence to C-MIT wear and number of activities carried out and time spent on them. Participants who did not have access to the Internet at home were supplied with an Internet dongle.

5.5.3.1 Encouraging Adherence

Participants were advised that research evidence shows that to get the most out of LifeCIT the C-MIT should be worn for 9 h a day, 5 days a week (i.e. a target of 15 days over 3 weeks). The website gave an option of between 1 and 9 h, and people were also advised to practise repetitive exercises and functional activities for 4 h a day. No specific guidance was given on the number of activities selected. However, patients were advised to ensure they had selected an appropriate number for their goals set at the start of each week. Data on adherence were self-reported via the website.

5.5.3.2 Method for the Phase II Exploratory Trial

An eight-centre, single-blinded pragmatic RCT compared usual care with LifeCIT. Participants were stratified by time post-stroke and into two ability groups based on the Orpington prognostic score. They were randomised to LifeCIT or usual care using computer-generated four-block randomisation. Group allocation was revealed after baseline assessment by an unblinded researcher opening sealed envelopes with the participant present. The researcher conducting the assessments was unaware of group allocation. A limitation of this was revealed when one patient referred to the intervention during the assessment. An additional measure to remove risk of bias was scoring of assessments by a second independent assessor from video recordings.

5.5.3.3 Participants

Two cohorts of participants were recruited: one was sub-acute participants following discharge from hospital, and the other was participants who were more than 16 weeks post-stroke. Those who did not meet the inclusion criteria at their initial assessment could be reassessed during the trial recruitment period.

5.5.3.4 Inclusion Criteria

1. Stroke (confirmed by clinical or neuroimaging diagnosis) affecting the upper limb
2. Either: (a) able to transfer safely between toilet, chair and standing, and able to walk safely at home wearing the C-MIT with or without the use of a walking aid, or (b) primarily a wheelchair user having help or supervision to transfer and walk

3. MMSE-2: Mini-Mental State Examination (MMSE), 2nd edition, score over 23
4. Minimum of 10° of active wrist extension and able to complete a grasp task from the Wolf Motor Function Test (WMFT)
5. Discharged home from hospital.

5.5.3.5 Exclusion Criteria

Co-morbidities that could interfere with participation: (a) severe pain of the hemiparetic shoulder, or (b) communication problems that could not be overcome by a carer.

5.5.3.6 Outcome Measures

Because this was an exploratory trial we used a wide range of outcome measures. These were: the MAL—28 (Uswatte et al. 2006), a patient-reported measure of upper limb activity subdivided into two sections: 'quality of use' (QOU) and 'amount of use' (AOU); upper limb section of the Fugl-Meyer (Fugl-Meyer et al. 1975); WMFT (Functional Assessment Score—FAS) (Wolf et al. 2001); stroke-related quality of life via the Stroke Impact Scale (SIS) (Duncan et al. 1999); and EQ-5D-5L (EuroQol Group 1990). In addition we used the Canadian Occupational Performance Measure (COPM) (Cup et al. 2003), a patient-focused and self-reported goal-orientated measure, subdivided into 'ability to perform tasks' and 'satisfaction with performance of tasks'. Tasks were set by the participant during an initial interview and were therefore relevant to their personal goals. Use of the LifeCIT website was recorded automatically by the LifeCIT software and generated self-reported data on compliance with the protocol in terms of the duration of C-MIT wear, the time spent on activities and the number of activities performed each day. Within 14 days of completing the intervention participants randomised to LifeCIT were interviewed about their experiences of using LifeCIT and taking part in the trial. Data generated by these interviews is now under analysis and will be published separately.

5.5.3.7 Assessments

Participants were assessed at home or in hospital at baseline (no more than 3 days before beginning the intervention), 1 week after completing the intervention and 6 months post randomisation. A researcher, blinded to group allocation, conducted all assessments except the post-intervention interviews. Adverse events and protocol deviations were documented. Standardised case record forms were used and double data entry into SPSS (Version 21, IBM Corp. Released 2012, Armonk, NY). WMFT and FM were recorded on video and scored at the end of the study by an independent blinded assessor to ensure consistency between different face-to-face assessors.

5.5.3.8   Intervention

The unblinded researcher opened a sealed envelope after the baseline assessment with the participant present. If the allocation was to the LifeCIT group the researcher gave the participant the Internet link to the LifeCIT website and a

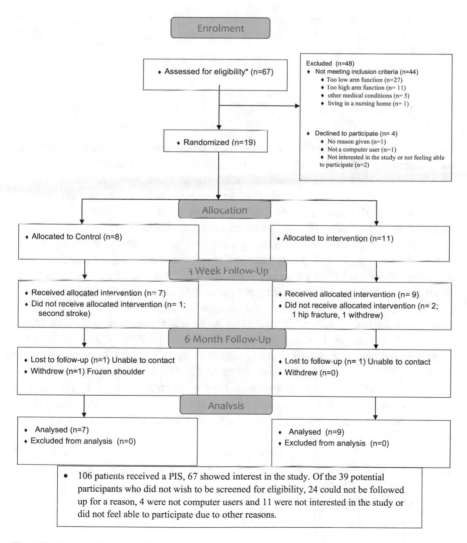

Fig. 5.8 Consort flow chart for research trial

C-MIT. For those who needed assistance with the computer, the researcher created a shortcut link on the computer desktop or showed a family member how to access the website. There was no further face-to-face contact during the intervention period. All participants continued with their usual care, which was documented by the unblinded researcher. If the allocation was to the control group the participant was asked to continue their normal care and an appointment was made for a 3-week assessment with a blinded researcher.

5.5.3.9 Data and Statistical Analysis

Recruitment, randomisation and adherence data are reported in accordance with the CONSORT (Consolidated Standards of Reporting Trials) (Schulz et al. 2011) (Fig. 5.8).

5.6 Results

From the 82 people with stroke who were sequentially screened, three were unable to use a computer and so declined to take part. Nineteen people satisfied the selection criteria for the study, gave written informed consent and were recruited: 11 were randomly allocated to the intervention group and eight to the control group. Importantly, there were no adverse or serious adverse events related to the intervention. Full details of the study and results are available in Meagher et al. (2016).

The intervention was well accepted and improvement in upper limb function was observed (see Fig. 5.9). A greater improvement was seen in the treatment group compared with the control group in all measures at each point that was statistically significant at the end of the treatment period but not at the 6-month follow-up. Minimum Clinically Important between group differences (MCID) were found in all measures at the end the treatment period and in the WMFT (FAS) at 6 months. MAL: 10 % (i.e. 0.5) (van der Lee et al. 2004), WMFT (FAS) 0.2–0.4 (Lin et al. 2009).

Adherence was monitored. Participants had been advised to aim for up to 9 hours C-MIT wear a day; self-reported mean C-MIT wear time was 4.8 (SD 2.6) hours a day, and activities were performed for an average of 3.2 (SD 1.7) hours a day. Participants had also been advised to wear the C-MIT and use the website 5 days a week, i.e. a target of 15 days; self-reported C-MIT and web-site use was 13.6 (SD 2.1) days.

In post-intervention independently conducted interviews, the participants reported positive effects on self-efficacy, confidence to use the affected arm and body image. Many participants said that they would have liked to continue using the system for longer.

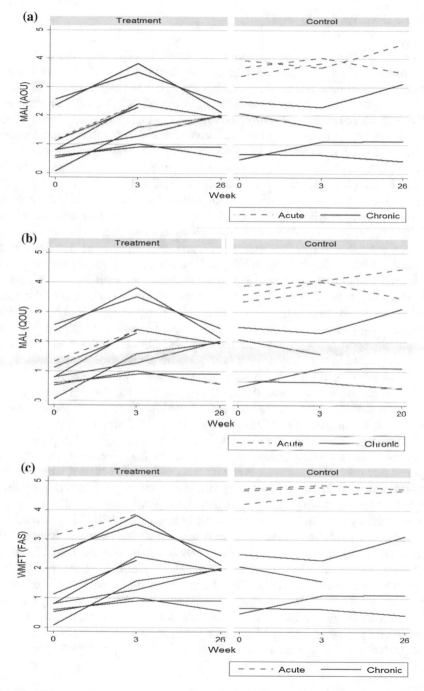

Fig. 5.9 Main outcome measures by treatment and control groups. Assessment scores at: Baseline (wk0), post-treatment (wk3) and follow-up (wk26): **a** WMFT (FAS score), **b** MAL amount of use, **c** MAL quality of use

5.6.1 Future for LifeCIT

This phase II exploratory trial compared two groups: LifeCIT and usual care. It was not powered to provide a definitive test of whether the LifeCIT intervention was superior to usual care, neither did it address the individual effects of C-MIT or use of the website alone, for which a different design (2 × 2 factorial or three-arm RCT) will be required. The trial demonstrated the potential effectiveness of the intervention and provided required data concerning the feasibility and variance estimates needed to determine sample size for a two and three-arm and 2 × 2 factorial design phase III trial. This data has enabled us to develop a feasible protocol for a phase III, definitive, multi-centre trial of LifeCIT.

Recruitment to a phase III trial is an important consideration. LifeCIT requires minimal physiotherapist training, which, as well as facilitating translation into clinical practice, will enable a study to recruit from a large number of clinical sites without excessive costs. Research nurses and therapists, with a minimum of training, can administer both the WMFT and the MAL, and if recorded on video, as in the exploratory trial, can be moderated to eliminate or correct for bias. We found minimal differences in scoring between assessors. Not only would this enable more participants to be recruited, it would also deliver a pragmatic trial in a clinical setting. An obvious limitation of the study was the dependence on patient-self-reporting, with potential inaccuracies. This limitation could be overcome by embedding a simple movement/pressure/temperature sensor within the C-MIT to record when it is being worn and the amount of movement.

At our dissemination event, therapists commented that the website was very useful as a motivational tool, but both patients and healthcare professionals voiced concerns about the patient's willingness to use the C-MIT protocol, especially early after discharge when they were coping with, and adjusting to, the problems of living at home following stroke. At this stage in their recovery, patients also reported that they expected to recover with time and it was not until later that they realised effort would be needed to regain use of their upper limb. This confirms previous research findings that some patients believe they are able to 'await' recovery during the early stages after stroke (Maclean et al. 2000). Based on this feedback web pages have been developed to support conventional therapy. This will allow, in the next study, a comparison between standard care, LifeCIT and the web-based aspect of LifeCIT without the C-MIT element. If the non-C-MIT, web-based system is found to be effective by comparing it with standard care, it could be used with a larger group of patients—i.e. those who lack sufficient movement to safely use the C-MIT. With additional web pages, it could also be used to support other aspects of stroke rehabilitation and rehabilitation of other neurological conditions such as traumatic brain injury or spinal cord injury.

5.7   Discussion

Improving the effectiveness of arm and hand rehabilitation post-stroke is important for individuals, their families and wider society. This chapter has discussed the rationale for this, together with a brief outline of the mechanisms of recovery and justification for the role of technology. It has considered the challenges of designing, implementing and translating technologies for the home environment. Patients' and carers' and healthcare professionals' perceptions of existing technologies and what they believe is most important for rehabilitation technologies in the future were outlined and include ease of use, comfort and safety. The chapter focussed in more detail on one example of such technology within the home environment, which was developed to address some of the problems of translation with constraint-induced movement therapy.

Further work needs to be done. New research designs need to be considered; the RCT may not necessarily be the best model for rehabilitation at home. Mechanisms need to be investigated, so we know what works for whom and why. Designed technology needs to be as intuitive as possible to potential users, capable of delivering high intensity goal oriented tasks, which can be used safely in a home environment.

The rigorous and systematic co-design of the LifeCIT intervention, which included physiotherapists, occupational therapists, patients and carers, was considered a critical factor in its acceptance by participants and in the excellent compliance demonstrated in the trial. LifeCIT was designed to complement usual care and to be used by healthcare professionals with patients. From our experience it is critical that all stakeholders are engaged in the design of technologies, protocols for use and translation into rehabilitation whether the technologies are delivered in clinical practice or at home. To ensure that the technologies are used, dissemination of information and education regarding the existence and availability of devices, as well as novel means of ensuring access and funding, need to be considered. Additionally with the shift towards strokes in resource-poor countries, affordability is likely to become increasingly important, driving the types of technology being developed and translation into practice. Incorporating changes in healthcare professionals and engineering training, could allow innovative technologies to be developed, while considering all of these other factors.

References

Boake C, Noser EA, Ro T et al (2007) Constraint-induced movement therapy during early stroke rehabilitation. Neurorehabil Neural Repair 21(1):14–24. doi:10.1177/1545968306291858

Burns A, Burridge J, Pickering R (2007) Does the use of a constraint mitten to encourage use of the hemiplegic upper limb improve arm function in adults with subacute stroke? Clin Rehabil 21(10):895–904. doi:10.1177/0269215507079144

Carey JR, Kimberley TJ, Lewis SM et al (2002) Analysis of fMRI and finger tracking training in subjects with chronic stroke. Brain 125(Pt 4):773–788

Cup EH, Scholte op Reimer WJ, Thijssen MC et al. (2003) Reliability and validity of the canadian occupational performance measure in stroke patients. Clin Rehabil 17(4):402–409

Dromerick AW, Edwards DF, Hahn M (2000) Does the application of constraint-induced movement therapy during acute rehabilitation reduce arm impairment after ischemic stroke? Stroke 31(12):2984–2988

Duncan PW, Wallace D, Lai SM et al (1999) The stroke impact scale version 2.0. Evaluation of reliability, validity, and sensitivity to change. Stroke 30(10):2131–2140

Ericsson KA, Krampe RT, Tesch-Römer C (1993) The role of deliberate practice in the acquisition of expert performance, vol 100. Am Psychol Assoc, pp 363–406

EuroQol Group (1990) EuroQol–a new facility for the measurement of health-related quality of life. Health Policy 16(3):199–208

Fitzpatrick G, Balaam M, Axelrod L et al. (2010) Designing for rehabilitation at home. In: Gillian RH, Desney ST, Gillian H et al. (eds) Proceedings workshop on interactive systems in healthcare (WISH10), Atlanta April 11 2010, pp 49–52

Fugl-Meyer AR, Jääskö L, Leyman I et al (1975) The post-stroke hemiplegic patient. 1. a method for evaluation of physical performance. Scand J Rehabil Med 7(1):13–31

Hanlon RE (1996) Motor learning following unilateral stroke. Arch Phys Med Rehabil 77(8):811–815

Hughes AM, Freeman C, Banks T et al (2016) Using the tuning methodology to design the founding benchmark competences for a new academic professional field: the case of advanced rehabilitation technologies. The Tuning Journal for Higher Education 3:249–279

Hughes AM, Burridge J, Demain S et al. (2014) Translation of evidence-based assistive technologies into stroke rehabilitation: users' perceptions of the barriers and opportunities. BMC Health Serv Res 14(124). doi:10.1186/1472-6963-14-124

Intercollegiate Stroke Working Party (2016) National clinical guideline for stroke. Royal College of Physicians, London

Kimberley TJ, Samargia S, Moore LG et al (2010) Comparison of amounts and types of practice during rehabilitation for traumatic brain injury and stroke. J Rehabil Res Dev 47(9):851–862

Kleim JA, Barbay S, Nudo RJ (1998) Functional reorganization of the rat motor cortex following motor skill learning. J Neurophysio l80(6):3321–3325

Krakauer JW (2005) Arm function after stroke: from physiology to recovery. Semin Neurol 25(4):384–395. doi:10.1055/s-2005-923533

Lang CE, DeJong SL, Beebe JA (2009) Recovery of thumb and finger extension and its relation to grasp performance after stroke. J Neurophysiol 102(1):451–459. doi:10.1152/jn.91310.2008

Lin KC, Hsieh YW, Wu CY et al (2009) Minimal detectable change and clinically important difference of the wolf motor function test in stroke patients. Neurorehabil Neural Repair 23(5):429–434. doi:10.1177/1545968308331144

Maclean N, Pound P, Wolfe C et al (2000) Qualitative analysis of stroke patients' motivation for rehabilitation. BMJ 321(7268):1051–1054

Meagher C, Ewings S, Burridge, J et al. Feasibility Randomised Control Trial of LifeCIT, a web-based support programme for Constraint Induced Therapy (CIT) following stroke compared with usual care. Physical Therapy: in press

Miller EL, Murray L, Richards L et al (2010) Comprehensive overview of nursing and interdisciplinary rehabilitation care of the stroke patient: a scientific statement from the American Heart Association. Stroke 41(10):2402–2448. doi:10.1161/STR.0b013e3181e7512b

Mozaffarian D, Benjamin EJ, Go AS et al (2016) Executive summary: heart disease and stroke statistics—2016 update: a report from the American Heart Association. Circulation 133(4):447–454. doi:10.1161/CIR.0000000000000366

Murray CJL, Lopez AD (1997) Alternative projections of mortality and disability by cause 1990–2020: global burden of disease study. Lancet 349(9064):1498–1504

Nakayama H, Jorgensen HS, Raaschou HO et al (1994) The influence of age on stroke outcome. The Copenhagen Stroke Study. Stroke 25(4):808–813

Nichols M, Townsend N, Luengo-Fernandex P et al (2012) European Cardiovascular Disease Statistics 2012. Brussels, European Society of Cardiology, Sophia Antipolis, European Heart Network

Nudo RJ (2003) Adaptive plasticity in motor cortex: implications for rehabilitation after brain injury. J Rehabil Med 35(41 Suppl):7–10

Nudo RJ, Milliken GW, Jenkins WM et al (1996) Use-dependent alterations of movement representations in primary motor cortex of adult squirrel monkeys. J Neurosci 16(2):785–807

Plautz EJ, Milliken GW, Nudo RJ (2000) Effects of repetitive motor training on movement representations in adult squirrel monkeys: role of use versus learning. Neurobiol Learn Mem 74 (1):27–55. doi:10.1006/nlme.1999.3934

Ridding MC, Brouwer B, Miles TS et al (2000) Changes in muscle responses to stimulation of the motor cortex induced by peripheral nerve stimulation in human subjects. Exp Brain Res 131 (1):135–143

Schulz KF, Altman DG, Moher D et al (2011) CONSORT 2010 statement: updated guidelines for reporting parallel group randomised trials. Int J Surg 9:672–677

Teasell R, Foley N, Richardson M et al (2013) Evidence-based review of stroke rehabilitation. Available via DIALOG. http://www.ebrsr.com/sites/default/files/chapter7_outpatients_final_16ed.pdf. Accessed 27 Jul 2016

Taub E, Crago JE, Uswatte G (1998) Constraint-induced movement therapy: a new approach to treatment in physical rehabilitation. Rehabilitation Psychology 43(2):152–170. doi:10.1037/0090-5550.43.2.152

Taub E, Miller NE, Novack TA et al (1993) Technique to improve chronic motor deficit after stroke. Arch Phys Med Rehabil 74(4):347–354

Truelsen T, Heuschmann PU, Bonita R et al (2007) Standard method for developing stroke registers in low-income and middle-income countries: experiences from a feasibility study of a stepwise approach to stroke surveillance (STEPS Stroke). Lancet Neurol 6(2):134–139

Uswatte G, Taub E, Morris D et al (2006) The motor activity log-28: assessing daily use of the hemiparetic arm after stroke. Neurology 67(7):1189–1194. doi:10.1212/01.wnl.0000238164.90657.c2

van der Lee JH, Beckerman H, Knol DL et al (2004) Clinimetric properties of the motor activity log for the assessment of arm use in hemiparetic patients. Stroke 35(6):1410–1414, doi:10.1161/01.STR.0000126900.24964.7c

Winstein CJ, Stein J, Arena R et al (2016) Guidelines for adult stroke rehabilitation and recovery: a guideline for healthcare professionals from the American Heart Association/American Stroke Association. Stroke 47:e98–e169

Wolf SL, Winstein CJ, Miller JP et al (2006) Effect of constraint-induced movement therapy on upper extremity function 3 to 9 months after stroke: the EXCITE randomized clinical trial. JAMA 296(17):2095–2104. doi:10.1001/jama.296.17.2095

Wolf SL, Catlin PA, Ellis M et al (2001) Assessing wolf motor function test as outcome measure for research in patients after stroke. Stroke 32(7):1635–1639

Wu CY, Trombly CA, Lin KC et al (2000) A kinematic study of contextual effects on reaching performance in persons with and without stroke: influences of object availability. Arch Phys Med Rehabil 81(1):95–101

Chapter 6
Telemonitoring in Home Care: Creating the Potential for a Safer Life at Home

Natalie Jankowski, Laura Schönijahn and Michael Wahl

Abstract The demographic development shows that society is getting older and especially rural regions have infrastructural deficiencies in the field of medical care. New technologies in the form of telemonitoring can offer elderly people a needs-based rehabilitative care in their homes. This article presents pilot projects that show new perspectives for the rehabilitation process.

Keyword Communication and information technology · Telemonitoring · Rehabilitation · Elderly

6.1 Introduction

Promoting health also means enabling elderly people to live in a familiar social environment in their own homes. Modern technology can help to make this happen for an increasingly larger number of elderly people.

Most elderly people want to live in their own homes. More than 2.2 million people aged 65 and over require formal or informal care (Marburger 2006). Recent statistics from the German Federal Statistical Office show that more than two-thirds of patients living at home use informal and formal care (Rothgang et al. 2015). With age, a number of risk factors arise, which must be taken into account: for example, deteriorating physical and mental health, reduced mental performance, living alone, lack of a family network, decreasing social contacts in a person's local area, and lack of access to primary care and shopping options. Elderly people need to use various strategies to compensate for the decline in their physical and mental functions. Another important issue is safety. Falls and dizziness often occur in later life, leading to constraints in everyday life.

In recent years, the home has emerged as a third health care location. Home care has many advantages, such as keeping people in a familiar environment, lower costs

N. Jankowski (✉) · L. Schönijahn · M. Wahl
Humboldt University of Berlin (D), Berlin, Germany
e-mail: jankowna@hu-berlin.de

© The Author(s) 2017
I. Kollak (ed.), *Safe at Home with Assistive Technology*,
DOI 10.1007/978-3-319-42890-1_6

81

and a lower risk of patients contracting an infectious disease (Fachinger and Henke 2010). At home, patients remain physically, mentally and socially more active and are integrated into a familiar environment. Using information and communication technologies (ICT) can help to improve the home care environment. The use of these technologies can provide a greater sense of security for older people and their families. ICT can facilitate additional medical care from various sectors. It can also help people to deal with their disease, while improving self-management and increasing motivation. In rural areas in particular, innovative technologies can support older who live a long distance from their doctor. However, for ICT to work, certain conditions have to be fulfilled. As well as requiring an area-wide broadband network, ICT requires users to accept a technological solution. Lack of acceptance is mainly due to low user involvement in the product development process (Wessig 2011). Other barriers to consider include concerns about stigmatisation, being observed and violation of privacy.

This article shows the potential of telemonitoring systems to help in home care, as well as barriers to their use. It focusses on monitoring people with an increased risk of falling, and those with diabetes and heart disease. Section 2 provides an overview of the theoretical background for the use of telemonitoring in the context of ehealth and telemedicine. Section 3 presents selected telemonitoring projects in the areas of fall prevention, diabetes and heart disease.

6.2   Telemonitoring in Health Care

Telemedicine involves using ICT in medical applications. The term telemedicine was coined in the mid-1970s (Häcker et al. 2008). In this field, telematic and telecommunication instruments support interaction and communication between doctors and patients at a distance (Häcker et al. 2008; Müller et al. 2009; Klar and Pelikan 2009). Telemonitoring is part of telemedicine and involves continuous monitoring of relevant parameters. Meystre (2005) defines telemonitoring as 'the use of audio, video, and other telecommunications and electronic information processing technologies to monitor a patient's status at a distance'.

Telemonitoring involves monitoring patients at home to increase their disease-related safety. This includes monitoring various health indicators over different time periods, for example, changes in weight, blood pressure profile, heart beat, blood glucose levels and lung function. Some clinical parameters can be monitored continuously. In the best case scenario, the monitoring is done wirelessly, which allows the monitored person to remain mobile. Continuous documentation gives a clear picture of the monitored person's condition (Goetz 2011). Telemonitoring has the potential to improve treatment quality for chronically ill people. This provides increased benefit to patients and helps to sustain good, affordable health care for all (Braun 2011). Telemonitoring enables better support of patients with chronic diseases, as well as more effective nursing. This increases people's quality of life and allows them to live independently at home for longer.

The benefits of telemonitoring procedures and their growing acceptance are shown in various national and international studies (John 2015). A study of one health care insurance scheme in Germany compared patients with diabetes in two different settings. Ninety-four patients where supplied with telemonitoring, while 300 patients received their usual care. The results showed that the patients in the telemonitoring group benefited from more effective treatment that prevented hospital admissions and shortened hospital stays, which were reduced to 50 %. The risk of complications was significantly reduced, saving on treatment costs. In addition, the data show that telemonitoring shifts medical and nursing services from inpatient to outpatient care, which makes patients more likely to comply with telemonitoring. Telemonitoring can also help to detect false or inappropriate care (Braun 2011).

In a project conducted by the Technical University of Munich, a telemedical blood pressure monitor was used, which automatically collected and sent data to a mobile device. A group of patients with increased blood pressure used the device, which allowed the doctor to see their blood pressure development over time. Comparison of these data with a norm data set from a rehabilitation clinic showed that more than 40 % of hypertensive patients did not actually have hypertension and had received unnecessary anti-hypertensive treatment. This shows that proper monitoring can prompt better diagnosis and save treatment costs (Wolf et al. 2011).

Telemonitoring can be used in the area of fall prevention. Falls are a major problem for the elderly, and the consequences can mark a turning point in people's life and mobility levels. Falls can be prevented using a simple system. The person carries a credit card-sized device in a pocket, which continuously analyses their gait and determines if they are moving normal or are at risk of falling. A signal alerts the person when the risk of falling increases. Thus a fall prevention system can be created that does not involve major modifications to buildings (Wolf et al. 2011).

All the technological innovations in the fields of telemedicine and telemonitoring must be appropriate for the user's cognitive abilities. Innovative devices need to be user-friendly and easy to use. It is essential that device developers understand elderly people's cognitive processes and learning abilities. It's not a case of the user having to understand the system, but of the system having to understand the user (Wessig 2011).

6.3 Projects and Sample Applications

The following section provides an overview of selected projects and products for fall prevention, hypertension, heart failure and diabetes. The selection is based on knowledge of the state of practice in the area of telemonitoring, where work has been ongoing for a number of years. Observations and searches were carried out at conferences, trade fairs and workshops. In addition, a search of the literature search was conducted. The selection is intended to give an impression of the state of technological development and practice in telemonitoring. Where available, central publications are given to complete the project descriptions.

6.3.1 Telemonitoring for Fall Prevention

Falls are a major problem for elderly people. Current statistics show, that 30–60 % of independent people over 60 suffer from at least one fall a year (Granacher et al. 2014). Fall prevention is important for preserving mobility, autonomy and quality of life in later life. Some factors cause increased risk of falling, for example, muscle weakness, imbalance, cognitive disabilities, aconuresis and restricted vision (Nikolaus 2013). Falls have many causes. Environmental factors such as wet floors and changes in terrain can trigger falls. The consequences of falls are wide-ranging. Alongside the cost to the health service, there are consequences for the person concerned, which include restricted mobility and reduced quality of life (Granacher et al. 2014). Following a fall the most frequent injury is a fractured hip. This leaves people dependent on others for daily activities. Most people cannot remain at home after a fall (Funk and Pierobon 2007).

The aim of the European Community and Australian Government funded *iStoppFalls* project was to develop an AAL system to predict and prevent falls by monitoring mobility-related activities and other risk factors that lead to falls among elderly people living at home. In addition to continuous fall-risk monitoring, the ICT-based system is designed to provide individualised training exercises. It is therefore important to have discrete measuring technologies and adaptive assistive functions that enable elderly people to integrate the system easily into their daily life. The project was implemented under the direction of the University of Siegen in cooperation with university and industry partners in Europe and Australia in the period 2011–14.

The *iStoppFalls* system has different components in the areas of home-based sensor technologies, telemedicine and video games. The Senior Mobility Monitor (SMM) is an inertial sensor system that continuously monitors mobility in daily life to acquire information on the frequency, duration and type of motion undertaken, as well as information on balance function and muscle power. The device can be worn as a necklace and so doesn't restrict mobility. Continuous data acquisition allows trend analyses of potential fall-risk indicators to be carried out. Furthermore, information on the effects of exercise can be provided and feedback given to the training system to adjust exercises in the Kinect-based fall prevention exercise training game, Exergame. The Exergame training system enables elderly users actively to prevent falls. Data acquisition is carried out by unobtrusive sensing and biomechanical modelling with a Microsoft Kinect camera and optional heart rate data assessment. In addition, an iTV application provides all relevant data for individualised fall prediction and prevention to elderly users at home (Fig. 6.1).

The system was evaluated in an international, multi-centre, randomised control trial. One hundred and fifty-three people living in the community (65+) took part between January and October 2014 in Germany, Spain and Australia. A comparison was made between an intervention group (n = 78) and a control group. The control group (n = 75) received only an educational book about general health and fall prevention. Participants in the intervention group received the educational book and

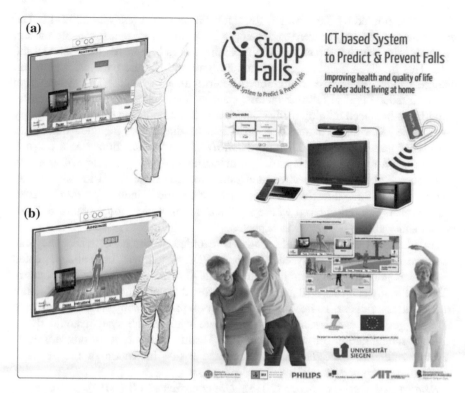

Fig. 6.1 iStoppFalls is a ICT based system to predict and prevent falls. © Copyright 2011–2014 iStoppFalls Consortium

undertook a 16-week exercise programme with individualised training sessions. During the intervention period, the median training duration was 11.7 h. The physiological risk of fall fell significantly more in the intervention group than in the control group. Furthermore, participants postural sway, stepping reaction and executive function increased with adherence to the training programme. The *iStoppFalls* system also received good ratings from participants for usability, user experience and acceptance. However, quite a low adherence rate—23 % of participants kept the weekly training programme—and gender differences regarding enjoyment of the system were found. One possible explanation is that Exergame wasn't challenging or enjoyable enough. Therefore, game design should take gender differences into account and create exercises that are sufficiently challenging. Furthermore, the design of the SMM device was reconsidered after men in particular said that they did not like wearing the device as a necklace.

Related publications: Vaziri et al. (2016), Gschwind et al. (2015), Wieching et al. (2012)

The Central European Institute of Technology investigated a wearable gait monitoring system that is integrated into pair of walking shoes. The *vitaliSHOE* project was funded by the Austrian Research Promotion Agency in 2009–11. The main goals of the project were to develop a multi-modal system for insole motion measurement, which indirectly monitors, maintains and improves the health of the user. This involves measuring, for example, gait and posture patterns, which can change rapidly, often due to unnoticed deterioration in health. The system also aimed to reduce typical fall-risk factors, such as immobility or reduced agility, by providing a balanced training method and long-term monitoring of physical activity levels. The latter allowed a motion profile to be drawn that included qualitative analysis of gait parameters. Thus the *vitaliSHOE* system functioned as a fall-risk detector and a motion analysis tool. The measurement system consists of a variety of sensors. They are mainly integrated into two shoe insoles and measure acceleration force in three directions and the angle of the foot's inclination on two axes. Thus, sensors are only placed where significant information about pressure distribution and walking behaviour is expected.

An additional wireless module sends all recorded data to a PC, where they can be stored, displayed and analysed. The *vitaliSHOE* system was, inter alia, validated in a clinical pilot study at a gait laboratory at the University for Applied Sciences in Vienna. Twenty subjects—10 patients who suffered from a singular proximal femur fracture after a fall and 10 healthy subjects—took part. All subjects had to undergo different geriatric assessment tests to evaluate whether the system could detect patients' rehabilitation progress after a fall. Healthy subjects were only tested to generate reference data. The results provided a first indication of the system's ability to capture and quantify patients' progress.

Related publications: Jagos (2013), Oberzaucher et al. (2010), Jagos et al. (2010)

6.3.2 Telemonitoring for Hypertension, Heart Failure and Cardiac Arrhythmia

Cardiovascular disease covers diseases of the heart and vascular system. In 2012, 50 % of all deaths in Germany were the result of cardiovascular disease (Vögele 2016). Well established treatments and effective drugs are available, but the most important factor in treating cardiovascular disease is patient compliance. The primary goal is to change the patient's life style. Drugs, stents and other treatments cannot replace intensive cooperation by the patient during treatment. Continuous blood pressure monitoring is necessary. High blood pressure causes various changes in the structure of the blood vessels. Since it doesn't cause pain, it is often recognised too late or is not taken seriously enough (Vögele 2016).

In the period 2011–13, the pharmaceutical company Daiichi Sankyo Germany worked in cooperation with the telehealth company almeda GmbH on a compliance project for patients suffering from hypertension. The project, named 'Blood pressure telemonitoring and telecoaching of hypertension patients in general medical practice',[1] examined whether additional monthly telephone support by health care professionals and telemetric blood pressure monitoring result in greater achievement of blood pressure goals. Bluetooth devices were used for data acquisition and transmission. After attending for 210 days, 96 patients showed significant reduction in systolic and diastolic blood pressure. Systolic blood pressure decreased significantly from 144.8 to 131.1 mmHg and diastolic blood pressure from 85.3 to 77.5 mmHg. Even higher reductions were found in patients with initially high systolic and diastolic blood pressure values. Additionally, patients' body mass index (BMI) declined during project participation. In the evaluable data for 61 patients, the average BMI decreased significantly from 31 to 30.5. In patients with an initially higher BMI (>30), the average value fell from 34.7 to 33.9. Moreover, 58 patients showed a significant increase in frequency of autonomous blood pressure measurement. The initial average measurement frequency of 'several times a month' rose on average to 'several times a week' and up to 'once a day'. Furthermore, patients with initially low measurement frequencies showed greater improvements. One possible explanation for the positive effects of telemonitoring and telecoaching is increase in patient awareness of their disease combined with greater treatment adherence.

Related publications: Ortius-Lechner et al. (2013)

The project *E.He.R.* ('Etablierung eines Versorgungskonzeptes für Patienten mit chronischer Herzinsuffizienz und Herzrhythmusstörungen in Rheinland-Pfalz') was launched to support regional, high-quality medical care for patients with heart failure or cardiac arrhythmia. It was initiated and funded by the Ministry of Labour, Social Affairs, Health and Demography in Rheinland-Pfalz between June 2012 and December 2014 within the project 'Health and Care—2020'. The use of telemonitoring was a key component of the *E.He.R.* care concept. During the project phase, 100 patients (mean age: 63.82; SD = 11.58) were supervised, of whom 68 % were seriously ill (NYHA II or worse). The patients lived up to 90 km away from the clinic they attended and were equipped with a personal scale and a blood pressure monitor. In addition some patients already had pacemakers. Measurements from all devices were transferred wirelessly to a transmission device in the patient's home and from there were automatically sent to a telemedicine centre. In the telemedicine centre, data were monitored 24 h a day, 7 days a week by medical experts. Patients and their doctors and cardiologists (even emergency doctors when required) were contacted regularly for an exchange of information by the telemedicine centre. If there were critical changes to the health of the patient, doctors and specialists could intervene quickly and effectively. In this way, each patient could be provided with individually adjusted therapy, and the care burden on doctors and specialists could also be eased.

After 6 months in the *E.He.R* care programme, decreases in typical symptoms were noted among patients: the percentage with severe breathlessness dropped from

---

[1]Original title: Blutdruck-Telemonitoring und Telecoaching von Hypertonie-Patienten in der allgemeinärztlichen Praxis.

48 to 12 %, and the percentage with relevant fatigue symptoms from 33 to 16.8 %. Physical limitations when climbing stairs or gardening were also less noticeable. Moreover, satisfaction levels with the care programme ranged from high to very high. Anxiety was reduced and about half of patients said telemedical care had a positive influence on their relationship with their doctor (no one cited a negative influence). For their part, doctors expected improved therapeutic options and better-quality patient contact. Based on the findings from *E.He.R*, a follow-up project, *EheR-versorgt*, was initiated in May 2015 and funded until April 2016 to extend the care concept to other regions.

Related links and publications: Zippel-Schultz et al. (2013), Helms (2015)

COMES (Cognitive Medical Systems) is a mobile diagnostic and therapy platform for patients with chronic diseases, such as diabetes, cardiovascular disorders, hypertension and heart failure. It was developed by the Heinz-Nixdorf-Chair for Medical Electronics at the Technical University of Munich in cooperation with Synergie Systems GmbH and Pasife GmbH. The aim of *COMES* is to enable medical diagnostics and intervention independent of time or location by measuring, transmitting and verifying biomedical data with established measuring techniques and communication structures.

COMES facilitates telecommunication between patients and physicians, mainly for preventative and clinical use and in rehabilitation. According to their specific needs, patients can choose among a range of *COMES* supported measurement tools to monitor, for example, blood pressure, blood glucose levels and even bruxism or breathing-related sleep disorders. Received data are automatically—and in accordance with data safety standards—transmitted via a smartphone or a tablet to the *COMES Trust Center*. They are then analysed and stored in a personal patient file, which is available for inspection by patients and doctors at any time. To add supplementary information to the file, which can be useful for data analysis and therapy, patients complete questionnaires on their smartphones. In addition to automatic data management, patients receive immediate feedback on current measurements on their smartphones or televisions. According to their preference, this feedback can come from a medical expert system or directly from their doctor. Thus doctors can provide personalised information, which leads to individualised therapy and follow-up.

Related product links: www.comes-care.net

6.3.3 Diabetes Mellitus

Diabetes mellitus is a metabolic disease that manifests in an increased concentration of blood sugar known as hyperglycemia. There are two types of Diabetes: type 1 and type 2. Both involve a reduced level of insulin. Treatment of diabetes focusses on avoiding dangerous diseases that can result from hyperglycemia and maintaining a good quality of life. The main point of the treatment is to prevent damage to the heart, blood vessels, kidneys, eyes and feet. Type 1 diabetes in particular requires active cooperation from the patient from the beginning. The missing insulin has to be injected using a syringe, a special pen or in some cases an insulin pump (Hodeck and Bahrmann 2014).

DREAMING (elDeRly-friEndly Alarm handling and MonitorING) received funding in the area of home monitoring from the European Commission in 2007–11. It was named best-practice-project within the scope of AAL at the final conference of the European Commission in June 2012. The aim of the project was to analyse existing treatment and care services for the elderly and to develop new standards of care with a focus on implementing innovative telematics infrastructure in six European countries (Denmark, Sweden, Germany, Estonia, Italy and Spain). Implementation was tested in a controlled follow-up study with 380 participants aged 60 years and over. Indications for participating patients were diabetes, chronic obstructive pulmonary disease (COPD) and chronic heart failure. Participants in all countries therefore received a home telemonitoring system provided by Health Insight Solution (HIS). The study integrated four applications:

- vital sensors to monitor health parameters such as weight, blood pressure, blood glucose levels and ECG readings
- environmental sensors for protection, such as smoke and movement detectors
- a mobile emergency call system with a fall detector and medication reminder
- a TV-based telecommunications system to improve social relations.

Data transmission could be static or mobile via Bluetooth. Wireless sensors sent data to a so-called HIS central unit in patients' homes. From there, the measurements received were automatically transferred to an HIS portal—a central, web-based telemonitoring platform on the HIS system. The HIS portal made the data available for inspection and further analysis by authorised people, such as patients and relatives. When selected individual limits were infringed, alerts were sent by email or SMS to the authorised people. Measurements were also transmitted to the coordination offices in the participant countries and periodically to the patients' GPs. Thus a practical information network was created, with the additional use of providing an integrated, electronic patient file. The results showed significant identification of previously unidentified diabetics, a 20 % reduction in falls and a drop of around 60 % in nursing home admissions. In addition, patients' quality of life improved, and reduced hospital admissions and shorter stays in hospital had a positive affect on health care economics.

Related publications: Giannakopoulos et al. (2009), Pflegewerk (2016)

VitaDock is a free software application developed and provided by Medisana AG. The app is designed for iPhones, iPads and MAC OS X, and enables storage, viewing and analysis of vital personal data, such as weight, body temperature activity, blood pressure and blood glucose levels. Users only have to connect their smartphone or tablet to a range of specific, Bluetooth-enabled devices provided by Medisana, for example, a body weight scale. From the app, data can then be transferred to the *VitaDock online* platform. *VitaDock online* offers a variety

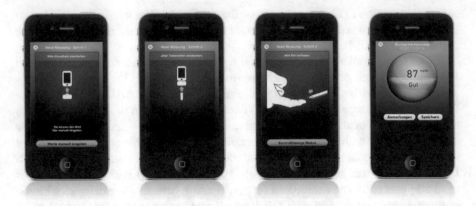

Fig. 6.2 Measuring blood glucose level with GlucoDock—4-step instruction. © Copyright 2011 MEDISANA AG, Deutschland

of data presentation and analysis options. The results can be synchronised on a tablet or smartphone using the app so the data are always within easy reach.

Two functions are presented below as examples. With the *GlucoDock* device, users can easily determine their blood sugar levels using their smartphone. *GlucoDock* has two adapters for measuring blood glucose levels: on one side, the iPhone/iPad adapter and on the opposite side, an adapter where the patient's blood glucose test strip is plugged in. Immediately after measurement with the test strip, the results are shown on the display and can be analysed and stored (Fig. 6.2). Additionally, the app allows the user to, for example, insert individual notes after each blood glucose test and to classify measured values with a view to setting individual target levels and analysing long-term data. As well as monitoring blood glucose levels, *VitaDock* offers an option to measure and manage blood pressure. When patients measure their blood pressure with the *CardioDock* cuff, the iPhone/iPad is plugged into the *CardioDock* docking station (Fig. 6.3).

Blood pressure data are transferred immediately to the mobile device where they can be viewed, analysed and stored. Individual settings and various analysis and presentation tools are also provided.

Related product links: http://www.vitadock.com

6.4　Outlook

Demographic developments and increases in chronic diseases in Germany mean health care delivery is going to change. In this context, home care will play a greater role (Naegele 2013). In addition, younger people will move away from structurally weak regions, changing the age mix in these areas (Flintrop 2009). People over 60 are now the largest age group in some cities and communities (Pachten et al. 2009;

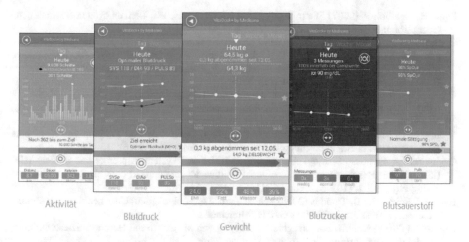

Fig. 6.3 VitaDock 2.0 App. Functions for activity, blood pressure, weight, blood glucose and blood oxygen. © Copyright 2016 MEDISANA AG, Deutschland

Ottensmeier and Jörg 2006). To guarantee adequate health care, new supply concepts have to be developed in the future (Lohmann 2009; Neumann et al. 2009). Because of these demographic developments, the cost of delivering health care and rehabilitation will increase further. In this article we have described how telemonitoring and other innovative technologies can be used to treat patients with chronic diseases.

Telemonitoring shifts ill people receiving medical and nursing care from inpatient to outpatient care, while at the same time boosting patient compliance with the treatment regime. The projects presented here show that elderly people with disabilities can enjoy greater safety through technology use. Psychoeducation can help chronically ill patients to cope with their disease, achieve better self-management, and become more motivated and proactive. Patients benefit from better treatment, better quality of life, fewer emergencies, fewer hospital stays and are able to stay in their own homes for longer (Fleck and Korb 2008).

References

Braun G (2011) Was ist für den Erfolg von Telemonitoring nötig? In: Picot A, Braun G (eds) Telemonitoring in Gesundheits-und Sozialsystemen. Springer-Verlag, Berlin, Heidelberg, pp 123–137

Fachinger U, Henke KD (2010) Der private Haushalt als Gesundheitsstandort. Theoretische und empirische Analysen, Nomos, Baden-Baden

Fleck J, Korb H (2008) Fernabfrage implantierter Defibrillatoren (ICD). Der Housecall-Telemedizin-Service von St. Jude Medical und PHTS Telemedizin. In: Jäckel A (ed) Telemedizinführer Deutschland. Minerva, Bad Nauheim, pp 151–154

Flintrop J (2009) Ärztemangel. Wenn der Nachwuchs fremdgeht. Dtsch Arztebl 106(09):396–397

Funk M, Pierobon A (2007) Sturzprävention bei älteren Menschen. Risiken-Folgen-Maßnahmen. Georg Thieme Verlag, Stuttgart

Giannakopoulos G, Greuèl M, Strohschein J (2009) Telemonitoring in den eigenen vier Wänden - das europäische Modellprojekt "Dreaming". Berlin. Available via DIALOG. http://www.pflegewerk.com/fileadmin/user_upload/Redakteure_Pflegewerk/Health_Care/GG_MG_JS_Dreaming_20091.pdf. Accessed 01 Aug 2016

Goetz CF (2011) Anforderungen der Nutzer an Telemonitoring. In: Picot A, Braun G (eds) Telemonitoring in Gesundheits-und Sozialsystemen. Springer, Berlin, Heidelberg, pp 39–49

Granacher U, Muehlbauer T, Gschwind YJ et al (2014) Diagnostik und Training von Kraft und Gleichgewicht zur Sturzprävention im Alter. Z Gerontol Geriatr 47(6):513–526

Gschwind YJ, Eichberg S, Ejupi A et al (2015) ICT-based system to predict and prevent falls (iStoppFalls): Results from an international multicenter randomized controlled trial. Eur Rev Aging Phys Act 12(10). doi:10.1186/s11556-015-0155-6

Häcker J, Reichwein B, Turad N (2008) Telemedizin. Markt, Strategien, Unternehmensbewertung. Oldenbourg Wissenschaftsverlag GmbH, München

Helms TM (2015) E.He.R. Für eine gute Versorgung von Menschen mit Herzinsuffizienz auf dem Land. Deutsche Stiftung für chronisch Kranke, Fürth

Hodeck K, Bahrmann A (eds) (2014) Pflegewissen Diabetes. Praxistipps für die Betreuung älterer Diabetes-Patienten. Springer-Verlag, Berlin, Heidelberg

Klar R, Pelikan E (2009) Stand, Möglichkeiten und Grenzen der Telemedizin in Deutschland. Bundesgesundheitsblatt Gesundheitsforschung Gesundheitsschutz 52(3):263–269. doi:10.1007/s00103-009-0787-7

Jagos H (2013) Therapy progress and gait stability monitoring in femur fracture patients via the wearable motion analysis system vitaliSHOE. Available via DIALOG. https://www.ige.tu-berlin.de/fileadmin/fg176/IGE_Printreihe/TAR_2013/abstract/Session-8-Event-1-Jagos.pdf. Accessed 27 Jul 2016

Jagos H, Oberzaucher J, Reichel M et al (2010) A multimodal approach for insole motion measurement and analysis. Procedia Engineering 2(2):3103–3108. doi:10.1016/j.proeng.2010.04.118

John M (2015) Telerehabilitation 2015. Medizinische Assistenzsysteme in der Prävention, Rehabilitation und Nachsorge. Available via DIALOG. https://www.fokus.fraunhofer.de/go/bericht. Accessed 15 Jul 2016

Lohmann H (2009) Gesundheitswirtschaft 2020. Vision einer Zukunftsbranche. In: Goldschmidt A, Hilbert J (eds) Gesundheitswirtschaft in Deutschland. Die Zukunftsbranche 1 WIKOM GmbH, Wegscheid, pp 732–742

Marburger H (2006) SGB XI-Soziale Pflegeversicherung. Textausgabe mit praxisorientierter Einführung. Walhalla Rechtshilfen, 6th edn. Walhalla und Praetoria Verlag GmbH & Co. KG, Regensburg

Meystre S (2005) The current state of telemonitoring: a comment on the literature. Telemed J E Health 11(1):63–69. doi:10.1089/tmj.2005.11.63

Müller A, Schwab JO, Oeff M et al (2009) Telemedizin in der Kardiologie 2010. Möglichkeiten und Perspektiven, Nucleus der Arbeitsgruppe Telemonitoring der Deutschen Gesellschaft für Kardiologie, Herz- und Kreislaufforschung. In: Duisberg F (ed) e-Health 2010. Informationstechnologien und Telematik im Gesundheitswesen. medical future Verlag, Solingen, pp 183–189

Naegele G (2013) Gesundheitliche Versorgung in einer alternden Gesellschaft. In: Hüther M, Naegele G (eds) Demografiepolitik. Springer Fachmedien, Wiesbaden

Neumann T, Biermann J, Neumann A et al (2009) Herzinsuffizienz: Häufigster Grund für Krankenhausaufenthalte. Medizinische und ökonomische Aspekte. Dtsch Arztebl 106(16):269–275

Nikolaus T (2013) Stürze und Folgen. In: Zeyfang A, Hagg-Grün U, Nikolaus T (eds) Basiswissen Medizin des Alterns und des alten Menschen, 2nd edn. Springer-Verlag, Berlin, Heidelberg, pp 113–127

Oberzaucher J, Jagos H, Zödl C et al (2010) Using a wearable insole gait analyzing system for automated mobility assessment for older people. In: Miesenberger K, Klaus J, Zagler W et al (eds) International Conference on Computers for Handicapped Persons. 12th International Conference, ICCHP 2010, Vienna, Austria, 14–16 July 2010, Proceedings, Part II. Springer-Verlag, Berlin Heidelberg, pp 600–603

Ortius-Lechner D, Wolf WP, Stumpe O et al (2013) Verbesserte Zielwerterreichung bei Hypertonie-Patienten durch ein Compliance-Projekt mit Fixkombinationen, Blutdruck-Telemonitoring und Telecoaching. Poster presented at 1. Bayerischer Tag der Telemedizin, Ingolstadt, 6 Mar 2013

Ottensmeier B, Jörg H (2006) Kommunale Seniorenpolitik. In: Bertelsmann Stiftung (ed) Wegweiser Demographischer Wandel 2020. Analysen und Handlungskonzepte für Städte und Gemeinden, Verlag Bertelsmann Stiftung, Gütersloh, pp 126–132

Pachten A, Reif S, Kunze H (2009) Potentialanalyse Telemedizin. Wirtschaftliche Wachstumschancen für die Medizintechnik in Berlin-Brandenburg. In: Jäckel A (ed) Telemedizinführer Deutschland. Bad Nauheim, Medizin Forum AG, pp 30–31

Rothgang H, Kalwitzki T, Müller R et al (2015) BARMER GEK Pflegereport 2015. Schwerpunktthema: Pflegen zu Hause, 36. Asgard-Verlagsservice GmbH, Siegburg

Vaziri DD, Aal K, Ogonowski C et al (2016) Exploring user experience and technology acceptance for a fall prevention system: Results from a randomized clinical trial and a living lab. Eur Rev Aging Phys Act 13(6). doi:10.1186/s11556-016-0165-z

Vögele C (2016) Herz-Kreislauf-Erkrankungen. In: Ehlert U (ed) Verhaltensmedizin. Springer-Verlag, Berlin, Heidelberg, pp 139–152

Wessig K (2011) Telemonitoring und Ambient Assisted Living: Anforderungen und Visionen. In: Picot A, Braun G (eds) Telemonitoring in Gesundheits-und Sozialsystemen. Springer-Verlag, Berlin, Heidelberg, pp 69–81

Wieching R, Kaartinen N, DeRosario H et al (2012) A new approach for personalized fall risk prediction & prevention: tailored exercises, unobtrusive sensing & advanced reasoning. J Aging Phys Act 20(8):120–124

Wolf B, Friedrich P, Spittler T et al (2011) Ambient Medicine®-Sensorik, Schnittstellen und Auswertung für telematische Diagnose und Therapie. In: Picot A, Braun G (eds) Telemonitoring in Gesundheits-und Sozialsystemen, Springer-Verlag, Berlin, Heidelberg, pp 83–109

Zippel-Schultz B, Budych K, Schöne A et al (2013) With Telemedical Networks towards a better Case and Care Management for Patients with Chronical Heart Failure and Rhythmic Disorders. In: Haas P, Semmler SC, Schug SH et al (eds) Nutzung, Nutzer, Nutzen von Telematik in der Gesundheitsversorgung—eine Standortbestimmung, Tagungsband der TELEMED 2013. TMF, Berlin

Online Sources

http://eherversorgt.de/
http://www.vitadock.com
http://www.comes-care.net
http://www.pflegewerk.com

Chapter 7
Empowering the Elderly and Promoting Active Ageing Through the Internet: The Benefit of e-inclusion Programmes

María Sánchez-Valle, Mónica Vinarás Abad
and Carmen Llorente-Barroso

7.1 Introduction

Sweeping changes occurred towards the end of the 20th century: technological developments linked to information and communications technology (ICT) and the Internet, and demographic changes caused by longer life expectancy. Both changes cause concern and merit study; they also create opportunities at the personal, socio-economic and political levels. According to the National Institute of Statistics, 13.6 % of the European Union (EU) population was aged between 65 and 79 in 2015; in Spain the figure is 12.6 % (INE 2016). Worldwide, between 2000 and 2050, the number of people over the age of 60 will double from 11 to 22 %. In absolute numbers, this age group will increase from 605 million to 2,000 million over half a century. Without doubt, 'there is no historical precedent' for this (WHO 2016).

According to a 2014 UN report, Spain will become the third 'oldest' country in the world in 2050, with 34.5 % of its population over the age of 65 (Aunión 2014). Life expectancy at birth is 81.6 years at present: 84.6 for women 78.5 for men. Some scholars consider this fact—demographic ageing—to be as important and remarkable as, for example, the Industrial Revolution was in its day, although it does not seem to receive as much attention as it should (Díaz 2015). Then again, elderly people today are not the same as they were a century ago: they have very good health, a higher level of education and, in many cases, high economic status. This leads to the conclusion that at present: 'Ageing is a development issue. Healthy older persons are a resource for their families, their communities and the economy' (WHO 2016).

M. Sánchez-Valle (✉) · M. Vinarás Abad · C. Llorente-Barroso
Department of Audiovisual Communication,
Publicity at the Universidad CEU San Pablo in Madrid (E), Madrid, Spain
e-mail: mvalle.fhum@ceu.es

© The Author(s) 2017
I. Kollak (ed.), *Safe at Home with Assistive Technology*,
DOI 10.1007/978-3-319-42890-1_7

This demographic situation—with all its political, economic and social conse-quences—occurs at a point in history that has also been compared to the Industrial Revolution: the advent of the Internet and ICT. Instant access to all kinds of readily available information and services has changed the world. It is undeniable that ICT can offer new opportunities to older people; this age group, whose options are sometimes limited by health or mobility issues, may find an ally in ICT. For this reason, research projects in this area are being carried out and encouraged. For example, the R&D report on ageing stresses the need for research into ways of combating the effects of human ageing through technology (Parapar et al. 2010). Our research shows that—in old age in particular—ICT provides ways of improving people's psychological processes, social aspects of their lives and issues related to dependency.

For this reason, the WHO (2002, p. 79) considers it is time to speak about active ageing as: 'the process of optimising opportunities for health, participation, and security in order to enhance quality of life as people age'. This concept is linked to physical, social and mental well-being, which includes older people's participation and integration in society (WHO 2002; OMS 2002). In view of the progressive ageing of the population, 'The challenge in the 21st century is to delay the onset of disability and ensure optimal quality of life for older people' (OMS 2001, p. 3). Thus, during the 1980s, the EU began to develop a new policy on ageing, which encourages a transition from a passive attitude to a more proactive one among older people. This new approach permits better welfare among the elderly and contributes to the economic sustainability of the EU's social systems, and so brings together the interests of all parties, including citizens, NGOs, businesses and political leaders (Walker 2009).

In this context, the Internet is a great support for active ageing. The percentage of older Internet users is expected to grow over the next few years, though growth will presumably be held back by access difficulties among those who are educated below secondary level. The digital divide is more evident among older people in modern societies. In that sense, the elderly make up the group most at risk of exclusion—or isolation (Fernández-García 2011). ICT provides ways through collaboration and by developing learning communities that overcome physical limitations, thereby providing opportunities for social integration and healthy direction (Agudo-Prado et al. 2013). Nowadays, older people need to learn how to use ICT and so reduce the digital divide between connected younger people and disconnected older people (Abad 2014).

Based on these data, we observe a digital divide that is generational and comes from different levels of access to and use of ICT in different social settings. Referring to the Internet, Castells (2011, p. 311) defined the digital divide as: 'the disparity between those who have and those who do not have Internet'. On the other hand, the Organisation for Economic Cooperation and Development (OECD) interprets the digital divide as: 'the gap between individuals, households, businesses and geographic areas at different socio-economic levels with regard to their opportunities to access information and communication technologies and to their use of the Internet for a wide variety of activities' (OECD 2001, p. 5). Only 68.5 % of older people have access to the Internet compared to 90 % of young people, and connection to the Internet drops after the age of 65 (Muñoz-Gallego et al. 2015).

This rupture between young and old, caused by unequal access to ICT, has become a major challenge for the UN and the European Commission. E-inclusion policies should, then, focus on helping the most excluded individuals to use ICT productively (European Commission 2007). In that sense, the EU Digital Agenda 2020 aims to make the most of ICT's potential 'to satisfy the needs of an ageing population' and so to contribute to active ageing.

In light of this, a research project entitled 'Digital Communication in Healthcare Institutions for Active Ageing' started at the University CEU San Pablo in 2012. Some of the research findings are presented here. The main objectives of the research were to understand and describe the opportunities that the Internet can offer older people in different areas, such as information, communication, electronic management and e-commerce. Particular emphasis was placed on health, as it is an important consideration for this group, and ICT can play a pivotal role in this area.

Information was obtained through group discussions, with quantitative data drawn from various secondary studies and reports. The wealth of information garnered during the conversations is presented here.

7.2 The Internet as a Tool to Empower the Elderly

Empowerment is a complex term (Rodríguez-Beltrán 2009). In general, empowerment is defined as a set of mechanisms that allow individuals to take control of their own lives (Silva and Martinez 2004). Zimmerman (1990) explains that empowerment means developing personal control with changes in self-perception, confidence, individual capacities and the skills needed to negotiate and influence decision-making (Rowlands 1997).

With regard to what the Internet can offer older people in terms of being independent and in control of their lives, the web provides many possibilities to empower older people. Indeed, Del Prete et al. (2013) show that the capacity to develop the skills needed to make rational use of the Internet becomes a stimulus for this group to continue learning. On this point, it is recognised that new information technologies (NIT) open new gateways for exchange, social participation and decision-making among the elderly (Del Prete et al. 2013). However, although the elderly appreciate the Internet's potential in different facets of their everyday life, they are relatively sceptical about the real role it plays in taking control of their lives (Llorente-Barroso et al. 2015).

Simply using the web shows a proactive attitude among older people, who see the Internet as a roundup of useful resources from many facets of their lives. This perception makes the web key to empowering older people, given that it allows them to take control over their lives due to its great potential.

7.2.1 The Internet as an Information Source for Older People

For the elderly, the Internet is one of the best sources of all kinds of information. With regard to the type of sites they use to search online, Google is their favourite search engine, and is the main gateway through which they access other sites in which they are interested. However, our research shows that this group tends not to use information platforms such as blogs and forums directly.

According to Miranda-De Larra (2004), older people are interested in the same subjects as other people; however, they also look for information that is relevant to the time of their lives. Our research uncovered the main subjects that interest them, which are ranked below in accordance with their importance to the elderly (Llorente-Barroso et al. 2015):

1. Current affairs: They show particular interest in online news that affects their geographical area (their town, province and country).
2. Health: This is a subject of special relevance to elderly Spaniards. They concentrate on searching for information that affects them more or less directly, such as illnesses, physicians, hospitals and healthy diets. The Observatorio Nacional de las Telecomunicaciones y la Sociedad de la Información (ONTSI) found uncertainty as to the trustworthiness of the information (54.4 %) and the risk of interpreting it wrongly (28.7 %) were the main obstacles (ONTSI 2012).
3. Culture and general interest: Older women, in particular, use the Internet to find recipes. In addition, the elderly seek information to satisfy their curiosity or to solve specific technical problems. In relation to culture, they tend to access information on exhibitions, travel, books, films, plays and other entertainment and leisure services (Fig. 7.1).

Generally speaking, uncertainty as to the trustworthiness of the information provided by the Internet is the main factor that threatens the web's potential to inform. The elderly are cautious and emphasise the importance of verifying

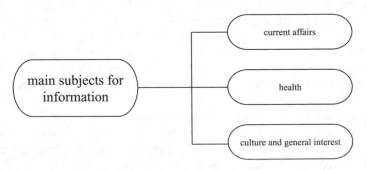

Fig. 7.1 Information from the internet—subjects older people are interested in

information and using reliable sources. The degree of caution varies according to the content, but everyone is particularly wary about healthcare information.

Apart from the source's reliability, other considerations that affect levels of confidence in the information offered include the design of the site and whether the platform has a good reputation.

Our research showed that the Internet's potential to provide information to the elderly promotes independent learning in this group and improves their well-being, as they practice new skills and increase their self-esteem. In this way, the Internet empowers older people by offering them an alternative source of information that gives them greater control over their lives.

7.2.2 How Older People Interact Online

Our research uncovered the following trends with regard to the Internet's potential to allow older people to interact.

1. In general, the means of online communication used most frequently by the elderly is email. For that reason, some of them favour smart phones because they provide immediate and convenient access to email.
2. Elderly people tend to limit their use of the computer to the means of communication on which they prefer to spend the most time. Thus, they emphasise media that they generally use to communicate with family members who live away from home, namely platforms such as Skype.
3. Increased use of mobile devices by older people has led them to interact more on social networks such as Whatsapp and Facebook. In fact, as is the case in other social groups, Whatsapp means they have fewer conversations on the landline. With regard to Facebook, older people see it as a social network for interacting with friends and family members. It is less immediate than Whatsapp but more enjoyable, as it allows them to share more expressive contents, such as pictures of their grandchildren. The reason this group of people belongs to a social network such as Facebook is because of their relationships with other people who are also on the platform and with whom they communication regularly. The fact that they use this platform and feel that managing other social profiles would waste time, means older people tend not to participate on other social networks such as Twitter. In any case, some elderly people have a negative opinion of social networks and see them merely as places to gossip where it is not worth wasting time (Fig. 7.2).

Our research found that the communication potential offered by the Internet acts as a catalyst for social interaction that integrates elderly people into social relationships with positive effects. On that respect, two positive effects stand out as the Internet fosters elderly people's motivation, self-esteem and satisfaction levels (Agudo-Prado et al. 2013):

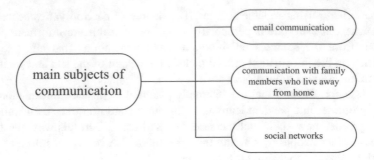

Fig. 7.2 Communication via internet—subjects older people are interested in

1. increase in social abilities
2. escape from isolation.

The communication potential offered by the Internet—and particularly the fact that it is mobile—allows the elderly to manage their social interactions online in a way that makes them feel good about themselves and in control of an essential aspect of their lives. Therefore, it is essential to consider this point when deciding how to empower older people through the web.

7.2.3 Older People's Relationship with Online Administrative and e-commerce Solutions

Online administration and e-commerce solutions are very helpful for people who have limited mobility or health problems that limit them in other ways (Miranda-De Larra 2004; Agudo-Prado et al. 2013). In that sense, the Internet facilitates some procedures for elderly people with such difficulties, as it allows them easily to carry out administrative and financial transactions from a distance. Our research found that, for this reason, the older generation finds an ally in the Internet thanks to e-commerce and online administrative solutions.

We found that the most important transactions carried out online by the elderly are as follows:

1. filing income tax returns,
2. managing bills and bank accounts and
3. making appointments.

Older people emphasise how easy and flexible it is to perform these tasks online compared to doing them in person in the traditional way.

However, the use of e-commerce is not as widespread among older people. In fact, according to our research, 'they see shopping online as unnecessary and complex', although 'a simpler shopping experience would increase online sales

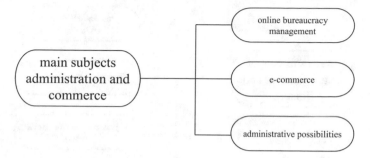

Fig. 7.3 Administration and commerce—subjects older people are interested in

amongst people over the age of 60'. We found that when they do use e-commerce, it is usually in order to:

1. buy train tickets,
2. buy cinema and theatre tickets and
3. book travel or hotels using platforms such as Trivago (Fig. 7.3).

Online administrative and commercial solutions allow older people to execute everyday activities more quickly. It is often a challenge for them to solve these kinds of problems online, but when they get used to it, they feel proud of themselves and satisfied. In this way, the Internet is not just convenient; it also allows them to solve problems and deal with tasks which some of them otherwise could not do because of their physical limitations. Therefore, the Internet makes them more independent and empowers them.

7.2.4 Leisure and Entertainment on the Internet for Older People

As mentioned above, the elderly use the web to find information about entertainment and leisure, but the Internet also them to consume culture. Thus, our research found that:

1. Several older people accessed radio and television programmes online, generally because they had missed the live broadcast.
2. Some older people were interested in fostering their cognitive activity and used the Internet as a sort of 'friend' to achieve this goal, as it offered them games that might improve their physical and psychosocial well-being (Blat et al. 2012).
3. Some older people used Spotify, although its use was not particularly widespread among the elderly.

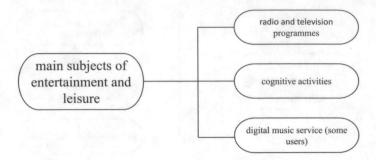

Fig. 7.4 Entertainment and leisure older people seek in the internet

In this sense, the Internet's creative and playful aspects allow for autotelic leisure (Cuenca-Cabeza 1995, 2004), where the priority is to carry out 'non-useful tasks' (Goytia-Prat and Lázaro-Fernández 2007) (Fig. 7.4).

Our research found that these online leisure outlets, while not the outlets most commonly chosen by the elderly from among all those available on the Internet, do increase elderly people's cognitive activity and self-esteem. They also give them control over their lives, which is proof of the Internet's potential to empower.

7.3 Health as the Main Concern for Older People When Surfing the Internet

The relationship between age and deteriorating health is obvious. The illnesses and complaints that are part of this stage of life mean people need to access more information about treatments, healthy living and medical services. The web is like a window that opens on to a great flow of information and it can assist older people in making more informed decisions about their health.

For older people, as for everyone else, the Internet has become an important tool for finding information about ailments which directly affect them or members of their families. According to a study carried out by the World Health Organisation, 80 % of Internet users look for information on illnesses or specific treatments, and 15–16 % consult doctor and hospital rankings. Online searches about health issues increase each year. In the United States alone, 18 % of Internet users have used the web for this purpose.

The e-health report produced by ONTSI (2012) shows that, among the sources most often used to obtain healthcare information, the Internet comes first after doctors, pharmacists, the print media and family. Of the searches carried out, 59 % are on private websites, although sites run by public institutions and medical journals are considered to be more trustworthy.

Older people are no different, and the majority of those who use the Internet repeatedly fall back on the web to find information about health issues, share their experiences with other Internet users and to carry out healthcare transactions.

The Internet has given the elderly greater control over their illnesses, symptoms and treatments. Used as an organisational tool, it allows older people to request medical appointments and receive reminders about appointments made without having to go to a health centre in person or wait for a telephone call.

The Internet's organisational use is complemented by information search options. Although medical professionals are still the main reference for older people when bothered by illness, they elderly find the Internet useful for understanding their ailments. Our research found that difficulty in understanding medical jargon is the starting point for many searches on the web after an appointment with a doctor. Internet searches, then, are seen as a complement to healthcare assistance and have great potential to become a source for a second opinion after a diagnosis.

Sharing experiences with other Internet users is another possibility opened up by the web, particularly in the case of people with reduced mobility. In this sense, the Internet has become a great showcase where healthcare professionals and health centres can offer their services. Although the elderly are not aware of the medium's commercial potential, they do use the web to compare healthcare services and to find information about their current doctor or a healthcare professional they may visit for a consultation.

At the same time, the abundance of information online about alternative therapies and complementary treatments appeals to older people and encourages them to use the Internet. For them the web is like a large medical encyclopaedia where they can find out about therapies and consult medical advice for a previously diagnosed health problem.

It may seem that older Internet users trust the web implicitly when it comes to investigating health problems, but traditional sources (doctors and pharmacists) continue to be their main point of reference, and access to digital information is not a substitute for relationships with healthcare professionals. However, speed and ease of access, and the fact information is available at any time, make the Internet a reference point that has more and more followers.

Regarding criteria for using the Internet, older people generally say they use good criteria to evaluate sites' trustworthiness (Sanders et al. 2015). They do not trust sites that look unreliable or do not provide documented sources. On the other hand, the design influences their perception of a website's quality as well as the contents. Sites that are quick to access and easy to use are used more often. Additional criteria that make sites more convincing are clear and accessible language and an attractive design (Llinas et al. 2005). This may seem like a good thing, but it has a negative side in that more appealing, user-friendly sites may inspire greater confidence than ones that offer more valuable content.

Another important problem for the older age group is difficulty accessing the Internet (Pernett et al. 2007), together with problems understanding online information, which sometimes requires good reading ability (Berland et al. 2001; Blanco and Gutiérrez 2002). The ONTSI (2012) found that one of the main barriers to using the Internet is uncertainty about whether the information obtained is reliable and fears about not interpreting it correctly.

Although older people find that the Internet allows them to access a great deal of information easily, the profusion of data is also seen as one of the medium's disadvantages because of the risk of misunderstanding the information obtained.

As for being able to find the resources needed to verify the reliability and accuracy of information, this requires a certain degree of technical skill. Real technical sophistication is sometimes required to use online resources. This is because of the enormous amount of information that can be found online, which complicates the selection of the proper sources. Older Internet-users mainly draw on Wikipedia and Google to consult the Internet, which is quite a basic way of going about it (Sanders et al. 2015).

On the same time, older people's socio-economic position favours using the Internet more to access healthcare information online. People who are more socially disadvantaged, who are often those more at risk of becoming ill, have less access to the web. It is probable that older people who have greater access to healthcare information online and are more able to obtain data on healthy lifestyle choices and on disease symptoms are more likely to visit a doctor because they can identify a possible health problem more promptly.

With regard to concerns older people have about accessing information about health issues online, they worry that they might become neurotic about their health. Ease of access to information may lead to disproportionate worry about ailments and illnesses and their symptoms, and people may consult online health sites too frequently and develop an unhealthy interest in the medium (Sanders et al. 2015).

7.4 Conclusions

From the results obtained, we may conclude that, in general, the Internet provides an opportunity to empower older people, to offer them active ageing that not only affects their own lives but also has an effect on society as a whole in all its economic, political and social dimensions.

The type of ageing that is future European societies' demands policies that stimulate independence in older people to promote sustainable and viable economies. In this respect, the Internet has become a tool that empower the elderly by offering them different ways to control their lives.

In summary, the web can empower older people in the following areas:

1. Information: our research shows that the Internet's potential to inform increases independent learning in this age group, helping to motivate them and promoting self-esteem. In this way, the web empowers the elderly by providing a complementary medium for obtaining information, which promotes cognitive activity among the older generation (Juncos et al. 2006, p. 185).

2. Communication: the Internet offers potential to communicate that gives the elderly control over their own social interactions. Our research found that online interaction gives them fundamental control over their social and civic lives, and this helps to combat potential isolation. This is a key reason to interpret the web as a tool for empowering the older generation.

3. Administration and commerce: our research found that the web allows older people to simplify administrative and commercial transactions. In this way, the Internet makes it easier and quicker to perform everyday tasks, which might otherwise not only be tedious due to traditional face-to-face procedures but actually impossible for some elderly people with physical limitations caused by health problems. Therefore, the Internet offers them independence, which helps to empower them.

4. Entertainment and leisure: our research found that online entertainment and leisure activities are not widely used or appreciated by the elderly, but these activities they do promote cognitive activity and self-esteem in this age group, thereby helping them to have control over their own lives (Fig. 7.5).

However, if older people are to exploit all the possibilities for empowerment offered by the Internet, digital literacy policies are required that will allow them to participate more fully in life at the civic level (Abad 2014). This proposal fits with the wishes of the elderly 'to continue exercising their rights as citizens and to participate in everything that concerns them and is important to them as members of society' (IMSERSO 2013, p. 16).

On this point, we consider it essential to propose e-inclusion programmes that permit the older generation to become more familiar with the Internet. These should have the added objective of allowing this group to make the most of the Internet to increase their independence and, consequently, the level of control they have over their own lives.

Secondly, assessments of the Internet as a source of healthcare information suggest it can play a positive role. Online healthcare communication helps older

Fig. 7.5 Older people and the Internet–main subjects of interest

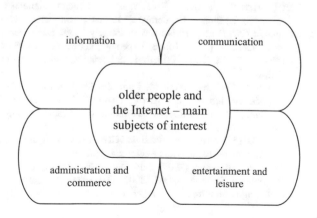

people to take charge of their own health and—when properly used—may, to a certain extent, work as a safety net that complements the work of healthcare professionals. Not having to rely only on the information received from their doctors gives older patients self-confidence and helps them to have more control over decision-making.

To sum up, the Internet provides an opportunity to empower older people and offer them active ageing, which helps both them and society. The digital divide continues to exist but is becoming more subtle, given that efforts made by the authorities to facilitate access and use are complemented by the interest the older generation itself shows in continuing to play an active role in society.

Acknowledgments As stated above, this chapter presents the results of the project funded by the University CEU San Pablo entitled 'Digital communication in healthcare institutions for active ageing', reference USPBSPPC03/2012, directed by Karen Sanders. The project's findings were first published as 'Internet and the Elderly: Enhancing Active Ageing' in *Comunicar* and in *Public Relations Review*, and we are grateful to them for the confidence they have placed in our work. In addition, we would like to thank Springer International Publishing and Ingrid Kollak for the opportunity to make these results more widely available.

References

Abad L (2014) Diseño de programas de e-inclusión para alfabetización mediática de personas mayores. Comunicar 42:173–180. doi:10.3916/C42-2014-17

Agudo-Prado S, Fombona-Cadavieco J, Pascual-Sevillano MA (2013) Ventajas de la incorporación de las TIC en el envejecimiento. RELATEC Revista Latinoamericana de Tecnología Educativa 12(2):131–142

Aunión JA (2014) No hay niños para el parque. El País 06-07- 2014. Available via DIALOG. http://goo.gl/b7uhm0. Accessed 14 Oct 2014

Berland GK, Elliott MN, Morales LS et al (2001) Health information on the internet: accessibility, quality, and readability in english and spanish. JAMA 285(20):2612–2621

Blanco PA, Gutiérrez CU (2002) Legibilidad de las páginas web sobre salud dirigidas a pacientes y lectores de la población general. Revista española de salud pública 76(4):321–331

Blat J, Arcos JL, Sayago S (2012) WorthPlay: Juegos digitales para un envejecimiento activo y saludable. Lychonos, Cuadernos de la Fundación General CSIC 8:16–21

Castells M (2011) La galaxia internet. Plaza y Janés, Barcelona

Cuenca-Cabeza M (1995) Temas de pedagogía del ocio. Universidad de Deusto, Bilbao

Cuenca-Cabeza M (2004) Pedagogía del ocio: modelos y propuestas. Universidad de Deusto, Bilbao

Del Prete A, Gisbert-Cervera M, Camacho Martí MM (2013) Las TIC como herramienta de empoderamiento para el colectivo de mujeres mayores. el caso de la comarca del montsià (cataluña). Pixel-Bit. Revista de Medios y Educación 43:37–50. doi:10.12795/pixelbit.2013.i43.03

Díaz AL (2015) Los mayores del siglo XXI: nuevas imágenes y nuevas perspectivas. Universidad de Mayores de Experiencia Recíproca, Madrid

European Commission (2007) European i2010 Initiative on eInclusion «To be Part of the Information Society». COM (2007) 694 Final. Available via DIALOG. http://ec.europa.eu/smart-regulation/impact/ia_carried_out/docs/ia_2007/sec_2007_1469_en.pdf. Accessed 18 Apr 2014

Goytia-Prat A, Lázaro-Fernández Y (eds) (2007) La experiencia de ocio y su relación con el envejecimiento activo. Bilbao: Instituto de Estudios de Ocio, Universidad de Deusto. Available via DIALOG. http://www.imserso.es/InterPresent1/groups/imserso/documents/binario/idi120_06udeusto.pdf. Accessed 22 Aug 2016

IMSERSO (2013) Envejecimiento Activo. Libro Blanco. Available via DIALOG. http://www.imserso.es/InterPresent2/groups/imserso/documents/binario/8088_8089libroblancoenv.pdf. Accessed 22 Aug 2016

INE (2016) Sistema Estadístico Europeo. Available via DIALOG. http://www.ine.es/ss/Satellite?L=es_ES&c=Page&cid=1254735905278&p=1254735905278&pagename=INE%2FINELayout. Accessed 25 May 2016

Juncos O, Pereiro AX, Facal D (2006) Lenguaje y comunicación. In: Triadó C, Villar F (eds) Psicología de la vejez. Alianza, Madrid, pp 169–189

Llinas G, Mira JJ, Pérez-Jover V et al (2005) En qué se fijan los internautas para seleccionar páginas web sanitarias. Revista de Calidad Asistencial 20(7):385–390

Llorente-Barroso C, Viñarás-Abad M, Sánchez-Valle M (2015) Internet and the elderly: enhancing active ageing. Comunicar 45:29–36. doi:10.3916/C45-2015-03

Miranda-De Larra R (ed) (2004) Los mayores en la sociedad de la información: situación actual y retos de futuro. Fundación AUNA, Madrid

Muñoz-Gallego PA, González-Benito O, Garrido-Morgado A (2015) Economía del envejecimiento. Centro virtual sobre el envejecimiento (Fundación General de la Universidad de Salamanca). Available via DIALOG. http://www.cvirtual.org/sites/default/files/site-uploads/docs/u28/file/web_estudio_economia_del_envejecimiento_.pdf

Economía del Envejecimiento. Nuevas tecnologías (2015) Universidad de Salamanca. Available via DIALOG. http://www.cvirtual.org/sites/default/files/site-uploads/project/u16/doc/web_nuevastecnologias_estudioeconomiaenvejecimiento.pdf. Accessed 15 May 2016

OECD (2001) Understanding the digital divide. Available via DIALOG. https://www.oecd.org/sti/1888451.pdf. Accessed 22 Aug 2016

OECD (2007) Working Party on Indicators for the Information Society. Available via DIALOG. www.oecd.org/sti/sci-tech/38217340.pdf. Accessed 23 Dec 2012

OMS (2001) El abrazo mundial. Campaña de la OMS por un envejecimiento activo. Available via DIALOG. http://www.who.int/ageing/publications/alc_elmanual.pdf. Accessed 12 Mar 2014

OMS (2002) Envejecimiento activo: un marco político. Revista Española de Geriatría y Gerontología 37(52):74–105

ONTSI (2012) Los Ciudadanos ante la E-sanidad. Available via DIALOG. http://goo.gl/EnTeh. Accessed 14 Jun 2014

Parapar C, Fernández-Nuevo JL, Rey J et al (2010) Report of R&D+i on Aging. I CSIC, Madrid. Available via DIALOG. http://goo.gl/PQwwDx. Accessed 26 Aug 2016

Pernett JJ, Gutiérrez JFG, Jiménez JLM et al (2007) Tendencias en el uso de Internet como fuente de información sobre salud. UOC Papers: revista sobre la sociedad del conocimiento (4). Available via DIALOG. http://www.uoc.edu/uocpapers/4/dt/esp/jimenez.pdf. Accessed 22 Aug 2016

Rodríguez-Beltrán M (2009) Empoderamiento y promoción de la salud. Red de Salud 14:20–31

Rowlands JO (1997) Questioning empowerment. Oxfam publication, Oxford, Working with Women in Honduras

Sanders K, Sánchez-Valle M, Viñarás M et al (2015) Do we trust and are we empowered by "Dr. Google"? older spaniards' uses and views of digital healthcare communication. Public Relations Review 41(5):794–800. doi:10.1016/j.pubrev.2015.06.015

Silva C, Martínez ML (2004) Empoderamiento: proceso. Nivel y contexto. Psykhe 13(2):29–39. doi:10.4067/S0718-22282004000200003

Walker A (2009) Active ageing in europe: policy discourses and initiatives. In: Cabrera M, Malanowski N (eds) Information and communication technologies for active ageing: opportunities and challenges for the europe union. IOS Press, Amsterdam, pp 35–48

Who (2002) Active ageing: a policy framework. World Health Organization, Geneva. Available via DIALOG. http://apps.who.int/iris/bitstream/10665/67215/1/WHO_NMH_NPH_02.8.pdf. Accessed 22 Aug 2016

Zimmerman M (1990) Taking action in empowerment research: on the distinction between individual and psychological conceptions. Am. J Community Psychol 18(1):169–177. doi:10. 1007/BF00922695

Chapter 8
Use and Development of New Technologies in Public Welfare Services: A User-Centred Approach Using Step by Step Communication for Problem Solving

Josef M. Huber, Helen Schneider, Verena Pfister and Barbara Steiner

Abstract Demographic changes and the resultant projected increase in older people needing care and support combined with a concurrent decrease in numbers of professional carers has led to increasing focus on possible technical care solutions. These solutions are supposed to support people who need care with their health, quality of life and level of independence on the one hand, and on other achieve cost savings in the health care system by keeping more people in their own homes rather than nursing homes. Against this background, this chapter shows contextual factors in this area in the first section, key drivers in the second and finally factors affecting success in the development and implementation of technical solutions. It takes account of the perspective of voluntary care services and focuses on lessons learnt in practice. In this way, this chapter provides important tips for implementing new projects—without going beyond the bounds of general knowledge.

Keyword Goals · Needs · Resource balance · Ethics · Side effects · Success factors

J.M. Huber (✉)
Evangelische Heimstiftung GmbH, Centre for Innovation,
Stuttgart, Germany
e-mail: j.huber@ev-heimstiftung.de

H. Schneider
University of Applied Sciences Koblenz - RheinAhrCampus,
Remagen, Germany

V. Pfister · B. Steiner
Bruderhaus Diakonie, Reutlingen, Germany

© The Author(s) 2017
I. Kollak (ed.), *Safe at Home with Assistive Technology*,
DOI 10.1007/978-3-319-42890-1_8

109

8.1 Contextual Factors in the Assistive Technology Area

Demographic changes and the resultant projected increase in older people needing care and support combined with a concurrent decrease in numbers of professional carers has led to increasing focus on possible technical care solutions (Hülsken-Giesler 2015). These solutions are supposed to support people who need care with their health, quality of life and level of independence on the one hand, and on other achieve cost savings in the health care system by keeping more people in their own homes rather than nursing homes (Hülsken-Giesler 2015). Against this background, this chapter shows contextual factors in this area in the first section, key drivers in the second and finally factors affecting success in the development and implementation of technical solutions. It takes account of the perspective of voluntary care services and focuses on lessons learnt in practice. In this way, this chapter provides important tips for implementing new projects—without going beyond the bounds of general knowledge.

8.1.1 Health, Illness, Disability

The World Health Organisation (WHO) takes the view that illnesses cannot be seen in isolation but unfailingly involve functional problems that entail additional loss of abilities. These can include, for example, restricted or impaired mobility, and an impaired ability to communicate, carry out day-to-day tasks at home, take care of oneself and interact with the outside world (Schuntermann 2006). For older people, this means that as well as compensating for age-related problems, such as hearing loss or being unsteady on their feet, they have to deal with restrictions in other areas, such as communication or social interaction. The challenge for support models, therefore, is not to alleviate individual problems but to provide an appropriate response to a medley of different factors.

A clear description of the functional problems makes it easier to perform a targeted intervention. The International Classification of Functioning, Disability and Health (ICF) (limiting information in: Bartholomeyczik et al. 2006) is useful for this. It delivers consistent terminology across all disciplines, allowing different service providers easily to understand one another and name limitations and daily routines. In addition, it covers different aspects of a person's functional health. Thus, according to Schuntermann (2006), a person is considered to be functionally healthy if, in the context of their overall, life:

1. their body functions and structure correspond to generally accepted norms. This includes emotional and mental functions (concept of body functions and structure)
2. they can do the things that are expected (concept of activity) from a person who docsn't have health problems (ICD—International Classification of Diseases)

Fig. 8.1 Activities as central factor in determining health problems. This representation of the model of disability that is the basis for ICF has been reprinted with permission of the World Health Organization (WHO), and all rights are reserved by the Organization (WHO 2002)

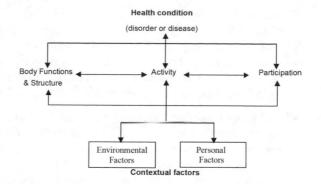

3. they can behave in all important areas of life as a person with no impairment would be expected to behave (concept of participation in areas of life).

The ICF model augments a purely biomedical perspective (concept of body functions and structure) with the idea of the individual as an active participant (concept of activity). The individual is viewed as an equal and autonomous participant in society and in their surroundings (concept of participation in areas of life). In this view, all external circumstances (environmental factors) and personal characteristics and attributes (personal factors) become important. For the purpose of the ICF, a person's state of functional health is a result of the interplay between functional-structural problems and contextual factors (Fig. 8.1). The paradigm shift achieved by the ICF's bio-psycho-social model is no longer to view functional problems as a personal attribute. Rather functional problems are the result of an interplay.

With this interdependent model, new perspectives on how to handle impairments to functional health open up. This means that solutions (and also undesired effects) can be sought not just by influencing body functions and structure but also in environmental factors, personal factors and in an individual's participation in their surroundings. Furthermore, each individual contextual factor influences the health problem and the remaining contextual factors. If one factor in the model changes, this will always effect the other factors, be that positively or negatively.

In the following chapter, these multi-factorial influences are outlined using the examples of selected technical solutions.

8.1.2   State of Development of Assistive Technologies

The study 'Supporting care-dependent people through technical assistance systems' provides a market overview of new systems on behalf of the German Federal Ministry of Health. The aim of this study is to identify assistance systems that improve the home care situation of care-dependent people or delay or prevent residential care. These technologies are supposed to be integrated into the catalogue of services offered by social care insurance in Germany at the same time.

Care-dependent people, their relatives and professional and informal carers were identified as the target groups. From 45 technical solutions, sometimes represented by several actual products, 12 technical solutions were identified that in principle are suitable to support home care and keeping people in their own homes for longer. After a cost-benefit analysis, six products were in the end identified that support living independently at home for longer (Weiß et al. 2013):

- *Toilet with wash function*: equipping the toilet bowl with a washing and drying function means visits to the toilet need no longer always be carried out in the presence of a carer. This greatly increases the independence of people cared for as outpatients. Even more importantly, it protects a person's privacy and sense of worth, and this can affect the relationship between the care-dependent person and the carer (Weiß et al. 2013). It simplifies self-care and in the ICF system, for example, addresses the activity of 'using the toilet'. Causal factor: structures and functions used in movement, for example arm movement, can be so altered by the ageing process that manual cleaning of the genital area is impeded.
- *Intelligent flooring*: A sensor mat can distinguish between people walking normally and falling. Using step recognition, orientation lights can, for example, be turned on. If several mats are used, it's possible to impute a fall if there is no activity for a long, unexplained period. Help can then be notified through an alarm system. Controlling lighting can reduce the risk of falls caused by not seeing things. After a fall, the time taken to find the person is reduced, preventing hypothermia and dehydration. Increasing the person's sense of security, counteracts anxiety as a factor in falls (Weiß et al. 2013). In the ICF system, for example, this addresses the activity of 'walking'. Causal factor: structures and functions used in movement, such as balance, sensitivity, circulation, and the musculoskeletal system, can be so altered by the ageing process that an increased risk of falling arises.
- *Electric medicine box*: the medicine dispenser provides reminders to ensure prescribed medicines are taken at the right time. In addition, connection to a telemedical centre means the situation is monitored by professional staff. This supports autonomy in administering the correct drugs, particularly for people on a number of drugs. The risk of taking too low or too high a dosage is reduced (Weiß et al. 2013). In the ICF system, for example, this addresses the activity of 'taking care of your health'. Causal factor: mental functions may not be sufficient to handle the demands of frequently administering precise doses or—as a personal factor—may deliberately not be focused on medication.
- *Automatic shut-off*: using a timer, the cooker is turned off after pre-set intervals. This reduces the risk of fire caused by forgetting to turn the cooker off, increasing the sense of security for care-dependent people and their relatives (Weiß et al. 2013). In the ICF system, for example, this addresses the activity of 'preparing meals'. Causal factor: mental functions are so focused on other things, due, for example, to stress, that the risk of fire caused by a forgotten cooker rises.
- *Mobile help standing up*: the process of standing up is supported through electromotive power reinforcement. This reduces the risk of a fall when standing

up and sitting down. Care-dependent people can move more independently, which reduces the burden on carers (Weiß et al. 2013). In the ICF system, for example, this addresses the activity of 'changing a basic body position'. Causal factor: structures and functions used in movement altered by the ageing process in the entire body.

- *Networked accommodation*: an information and communication system allows all kinds of facilities to be delivered through static and mobile touch displays, including telephone and video call services. Ease of making contact and interacting socially from a person's immediate surroundings is believed to make it easier for them to stay in their own home or to make lasting contacts quickly in their district after a move (Weiß et al. 2013). In the ICF system, for example, this addresses the activity of 'community life'. Causal factor: taken together physical limitations can lead directly or indirectly to changed social behaviour.

The study shows that both simple devices and complex systems that connect different service providers can contribute to improving care-dependent people's quality of life, independence and sense of security. With reference to the ICF's environmental factors, it also shows how interdependent and fragile the living environment of care-dependent people can be:

1. Products and technologies (for example, aids and medicines). Here, it can be seen that the availability of simple and reliable everyday aids, from wet wipes to walking frames, can raise the threshold for investment in a technical solution.
2. The natural and man-altered environment (for example, buildings, streets and footpaths). Existing infrastructure, such as hand showers, bidets, banisters and light switches that are easy to reach can qualify the additional functional benefit of a technical solution. At the same time, missing infrastructure, such as mains water, electric or telecommunications connections can prevent technical solutions from being deployed.
3. Support and relationships (for example, family, friends, employers, and health and social care experts). Here, it's been shown that technical solutions can relieve informal and professional carers. Resources can then be reallocated—with the risk that they may go away.
4. Attitudes, values and convictions of other people in society (for example, the economic system's attitude to part-time work). Beliefs and practices in the areas of hygiene, security, self-determination, welfare, consumption or social interaction can affect how relevant a functional or structural problem is considered to be.
5. Services, systems and operational principles (for example, the health and social system with its benefits, services and legislation). Financial constraints, building regulations and questions of legal liability mean an innovation may not be implemented for financial or legal reasons.

Reliable cost-benefit analyses are not possible at the moment, because only a few approaches are as yet ready for market and there are no large-scale studies available quantifying their benefits. To develop the area further, effectiveness studies, business models and information models should be implemented. Furthermore, ethical and legal issues must be clarified. From a social law perspective, independence, participation, promoting a sense of security and relieving carers can be used as arguments for financing. The prevention of care dependency can be re-imagined through these new options (Weiß et al. 2013).

It's beyond the limits of individual research projects to determine whether hoped for potential savings will be offset by follow-up or additional costs. We should remember that reductions in staff may mean additional expenses for nuisance alerts, maintenance and administration. Last but not least, we should reflect on the effect of (false) expectations from the technology on our understanding of values and behaviour (Manzeschke 2013).

8.2   Key Concepts for Problem Solving

If problems are to be solved in a sustainable way, it's important to ensure that the solution doesn't create new problems. This is particularly true of problems whose consequences for the individual or society are no better or even worse than the problem itself. In this respect, it's important to observe that solutions have their own momentum. The paramount objective of solutions must be to optimise balance in life and give everyone the greatest possible freedom. When targeting an objective, the cost-benefit equation must be balanced (Illich 1998). In such a cost-benefit equation, it is helpful to analyse the problem to be solved closely, to set considered goals and finally to weigh up the resources needed and potential outcomes of the measures. The paramount objective of 'giving everyone the greatest possible freedom' introduced by Illich already implies that this consideration should go beyond technical aspects to ethics.

8.2.1   Problems—Problem Analysis

Problem analysis is repeatedly used in day-to-day practice and in science to overcome problems strategically. For this, the problem should be described as exactly as possible. This involves identifying patterns of cause and effect connected with the problem (Elsbernd 2013). To recognise concrete problems, it's helpful to understand how they generally arise: problems arise when a targeted objective cannot be reached either because the necessary means are absent or because it is not known how to reach the objective. Problems are characterised by a (1) barrier or gap, which prevents a transformation from an (2) undesired initial state to a (3) desirable end state (Duncker 1966; Dörner 1987; Lüer and Spada 1992). When dealing with problems, the perception of the problem as well as its context must be considered (Lüer and Spada 1992). Unsolved problems in the sense of high maintenance are one of the conditions for care needs (Hasseler and Görres 2010).

Example:

In the context of counselling, a husband carer describes the practical relevance of looking at a problem in different ways: his wife, who has dementia, is complaining of pain in her abdomen (undesired initial state). She knows and seeks the state of pain-free well-being (desired end state). For the woman, the problem manifests as an unsatisfactory state of helplessness/powerlessness. She doesn't know how to achieve the state of being pain free (barrier). The husband has learnt how to resolve the situation. Both have a common interest in a state of well-being being reached. The husband has learnt that the painful stomach results in restive behaviour. However, he has also realised that the pain is the result of a full bladder. With this analysis of the problem, he can overcome the barrier that exists for his wife. He takes her to the toilet. There she realises what the problem is and can solve it herself.

Tip:

Question the problematic situation: what exactly is the problem? Who has the problem? Are other people involved? Why exactly has the problem occurred?

8.2.2   Problems—Problem Evaluation

For concept development in professional care, an evaluation of the problems to be tackled with the concept is vitally necessary (Elsbernd 2013). Otherwise, there is a danger of only solving problems 'in the test tube': instruments and procedures are used on their own or are not sufficiently aligned. Through various system archetypes, for example, problem transfer, this can lead to additional work or conflicts (Elsbernd 2013; Senge 2011). When evaluating problems, the relevance, dimensions and controllability of the problem and connected themes and problems should be reviewed. Relevance describes the consequences of not solving the problem from the perspective of different stakeholders. The dimensions of the problem refers to how deeply, for how long with what potential consequences the problem appears in everyday life. The controllability of the problem helps to broaden the gaze to different possible solutions in that it asks who can control the problem. Correlated problems and themes take account of areas and processes where a problem can take control and illuminate processes and structures that can prevent or support the problem being solved (Elsbernd 2013).

Example:

If we look at Uli Anders' problem as being the risk of falling in the dark, it can be analysed more closely. The risk itself has no consequences for Uli. However, his relatives are already worried about the risk—which may encourage caring behaviour in them. For Uli, the relevance is noticeable in the problem's dimensions: because is afraid of falling he doesn't go to the toilet although he has a full, bulging bladder. As a consequence, his sleep is affected. In the long term, his urogenital tract could be damaged. In terms of a single night, the problem lasts until the morning. Then it becomes light and Uli can go to the toilet without bumping into anything. In terms of the longer view, the problem persists. Then a disturbed night's sleep could lead to changed social behaviour. The worry about urgency in the night could prompt Uli not to drink in the evening.

In the BMBF Project MOVEMENZ (Mobile, self-determined living for people with dementia in the neighbourhood), technologies were to be identified that contribute to preserving mobility and movement in people with dementia. To that end, the whole care arrangements for people with dementia were examined to develop technical solutions in a demand-pull approach. With an iteration loop, the technology developers formulated suggestions for solutions to the established problem areas. The project results were characterised by different problem perceptions at all levels (Weinberger and Decker 2015). Even in connection with the question of acceptance of technologies and assistance systems, the MOVEMENZ project showed how important it is to analyse the problems at hand accurately and in their context. In the example of a tendency to run away, it becomes clear that even here different problem perceptions exist. A tendency to run away is connected with the risk of loss of orientation. That can lead to false assessment of a situation or helplessness. Connected with that is the risk—sometimes fatal—of an accident or hypothermia. It's difficult to show this risk statistically (see the section 'Risk and security'). While the risk itself has no relevance for people with dementia, it's of great importance to carers and relatives and leads—possibly conditioned by a media-distorted perception of the actual, numerical, recordable risk or by focusing on worst-case scenarios—repeatedly to worry (for example, reproaches and loss of control) and concern (for example about the person with dementia or third parties). Thus a risk repeatedly becomes the tangible dimension of diffuse fears. This is also true for the affected people themselves when they try to return to their own four walls because of worry, stigmatisation or disorientation. So the dimensions of the problem lie on the one hand in an accident happening. The frequency, duration and consequences of the risk can be estimated here on a case-by-case basis. On the other hand, changes to everyday freedom and participation in social life must be considered. In this case, the actual handling of the risk must be assessed: should structures be put in place in daily life that reduce or prevent the occurrence of the risk—or should structures be created that alleviate the consequences that arise when the risk occurs? Movement itself has an influence on well-being, cognition, sleep, mood and the cardiovascular system (Andrews 2012a, b; Baillie et al. 2012). The correlated themes and problems clearly demonstrate how diverse an isolated problem, such as a tendency to run away, can be. The structures and processes of the (health) systems in the western world are oriented to economic efficiency. The opportunities to correct mistakes are therefore small. This increases the demands on all who play a part in the system—including service recipients, such as people with dementia. Symptoms, causes and triggers of mistakes often aren't perceived as being different as the question of 'blame' is raised. At the end of this chains stands a person with dementia who is often perceived as problematic or even as a problem. If the symptom (malfunctioning of the system) is confused with the cause (in-creased susceptibility to malfunctioning) and guilt assigned to the trigger (for example, people with dementia who act or employees who react), it can lead to problem displacement and escalation (Senge 2011): Challenging behaviour, Specialling, Forcible confinement, Monitoring.

Tip:
    Explore the context for a problem: what exactly is the current initial state? Are there structures and processes that are influencing the problem. How do other stakeholders answer these questions?

8.2.3 Goals

Dörner divides problems into four types, according to whether the level of awareness of the solution is high or low and the goal criteria are clear or less clear (Dörner 1987). Goal criteria can be understood as reduced but sufficient descrip-tions of goals. In practice, the description of goals appears complex. However, formulations such as SMART (specific, measurable, achievable, relevant, time-bound), RUMBA (relevant, understandable, measurable, behavioural, achievable/attainable) or DREAM (distinct, relevant, evidence-based, achievable, measurable) support this description (Koutoukidis et al. 2013; Schiemann and Moers 2004). Goals should be viewed as mental anticipation of a desired end position. Explicit goals can be expressed in conversation or on request. Implicit goals, by contrast, are hard to recognise. Different players in the same situation can pursue complementary or contradictory goals (Röhner and Schütz 2016). Alongside problems, goals are a further factor influencing care needs. For the individual goal of daily personal hygiene, carers could set the care goal of obtaining self-reliance or of passively compensating limitations. In goal setting, professional, cultural, soci-etal and social values and norms come into play (Hasseler and Görres 2010). In particular, individual moral values and mental models mean that misunderstandings

Fig. 8.2 Levels of deviation
in goal determining

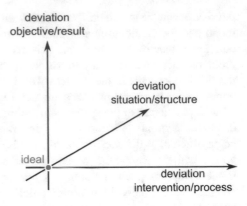

occur on various axes. Therefore goals can only ever be reached with limited accuracy. Figure 8.2 shows the axes that determine goal deviation.

Goals can be misleadingly formulated and outcome indicators can be unclear. Situations can deviate from mental models and structures may not correspond to actual requirements. Last but not least, measures or processes implemented can be variously understood and a goal may thus not be reached.

Example:
   In care counselling, Uli Anders can agree a primary goal for his problem 'risk of falling in the dark'. The goal can be oriented to compensation, rehabilitation or prevention. This overarching objective decides and justifies if and why resources are maintained or promoted. If necessary, it also justifies why removal of resources must be approved.

The BMBF project QuartrBack (intelligent emergency chains for people with dementia) is supposed to enable people with dementia to have more mobility, independence and participation in social life. People affected are supposed to be able use a technical device to request support from volunteer or professional helpers (www.quartrback.de). During the project, it became ever more clear through the user-integrated approach and deepening research that the shape of daily life for people with dementia needed to be acknowledged in the QuartrBack system. For the structures, processes and goals that an emergency chain must connect to are anchored in daily life. For everyone involved that means using the technology in daily life with added value, so that it can be used in an emergency. Experience with home emergency call devices was transferred for this. The label 'emergency button' acts as a barrier here. This is partly because the subjective perception can dominate that the predicament isn't an 'emergency' (for example, underestimation, trivialisation, shame, anxiety about consequences) and so the button deliberately isn't pressed. Sometimes the emergency button isn't worn on the body—sometimes so as not to damage it, to keep it for an emergency or because it's annoying to wear— and can't be reached in case of need. This is also a reason why use of the emergency button needs to be practised to overcome inhibitions. This can also answer practical questions: which arm does the emergency button feel more comfortable on? Would it be better worn round the neck? Is communication with the home emergency call centre possible from all rooms? Even if the radio or TV is on?

Furthermore, through problem analysis and problem evaluation, the shape of daily life was revalued by placing the problems to be solved in the context of daily life. Thus the idea of designating railway tracks and rivers as dangerous zones was scrapped. It is technically doable to estimate current hazard potential. However, having a helper at the right place at the time of the emergency requires this helper to be alerted in good time. Timely alerts increase the number of false alarms and lead to lower acceptance of the system through to collapse of the helper network. This reflection shifts the needs and concerns of the helper to the foreground. An essential element in QuartrBack is therefore the support of established offerings for people with dementia in their local area or the promotion of new offerings. Networking all players is now of particular significance.

Dialogue on an equal footing is essential for this process, as is appreciation for the expertise of all those involved and above all constant effort to metabolise and understand the perspectives of others. In this process, it's challenging to exchange different interests and moral values, to speak about conflicting values and find a consensus and a common way forward. Appreciation of the role of society and professional service providers with regard to technical innovation doesn't just lead to a successful balance of interests. The supposed or relative 'devaluation' of the importance of technology leads to ascertainment of responsibilities and demands. This puts a brake on excessive expectations of the technology—we considered sensitivity to the limits of technology to be an important success factor and a counter measure to the idea of omnipotent technology.

> Tip:
>   People's objectives vary even more from person to person that their perception of problems. In inter- and transdisciplinary collaboration in particular, it's helpful to communicate governing values and norms with trust, as well as to communicate unease. Don't just send a signal that voicing concerns is allowed. Make discussing these subjects a duty.

8.2.4   Resources

With his division of problems, Dörner doesn't just address the clarity of goal criteria, but also the awareness level of the means to a solution (Dörner 1987).

Resources are the abilities, skills and influences that are at a person's disposal to reach his demands, that is his (potential) goals. Resources can be bound to a person or a person's body as physical, mental or spiritual resources (for example, statics, motor skills, sensors, cognition, feelings, self-confidence, resilience, creativity). They can also be unleashed by him, for example spatial (for example, living space or living environment), social (friends, relatives, social network, state), economic (for example, money, possessions) or spiritual resources (for example, belief, piety).

They can be available, perceived, activated and developed as material and as intra- and interpersonal quality and they can be built up and they can be degraded (Hölzle 2011; Thieme 2015; Zwingmann 2007; Mahler 2012).

Example:

By clarifying resources, it can be seen that the problem only predominates at night. Only at night does worry about falling stop Uli from emptying his bladder. With regard to walking, Uli Anders has adequate physical and mental resources, including cognition and motivation.

In the BMBF project QuartrBack introduced above, objectives were reflected in a picture of separate functions. Technology takes over the sensors' functions. It oversees critical parameters, evaluates them and sends alarms to the Service Centre Care. The people in the Service Centre Care analyse and assess the situation. They decide and are responsible for the help and support measures that are implemented by the helper network in a help scenario. In the final analysis, the people in the helper network are the actors. As executive forces, they look for the place of the happening in order to help the helpless person. Problem analysis, problem assessment and definition of objectives extended the work area. A significant part of the additional outlay that is necessary for lasting implementation of the concept could thereby be discerned early on and included in work plans. Considering carers' and relatives' existing strategies for handling existing problems limits the above mentioned additional outlay as established concepts and experiences can always be built upon.

Inclusion of existing strategies can also cause a change in the way of thinking. In severely restricted people with a deficit-oriented perception, the question of which everyday objects can house a geolocation and communication system led to arm and foot bindings with a trick lock and aids with a dependency relationship such as rollators and walking sticks. The reverse question formulation with a question about resources brought additional options: 'Which communication medium can be located and has attractive functions for daily life?' This question formulation opened the eyes and broadened the gaze to owning a smartphone. Equipping a Smartphone with additional features doesn't just draw on the resources of the people in the QuartrBack system. It also brings additional resources into play in project development. Smartphone-based experiences can be used as a basis for developing independent devices.

> Tip:
>   Analyse players' existing solutions. Use the experience and knowledge of these experts.

8.2.5   Needs

Needs are the gap between the resources necessary to reach a goal (desired situation) and the resources available to a subject (current situation). This need also expressly includes perception and knowledge deficits. This can be relevant when qualitatively and quantitatively all the resources needed to reach a goal are available but are not adequately orchestrated actually to reach the goal (Fig. 8.3). Needs are also discussed when a goal cannot be reached or can only partially be reached. A care need exists when the goal in question is a care goal. If the care goal has been determined by third parties, such as the care profession, the need can be defined as an allocation-ethical need. Thus it is afflicted with the risk that individual needs will be overlooked and nominating, standardised needs formulated. This risk must be counteracted with participation concepts such as Shared Decision Making. Needs

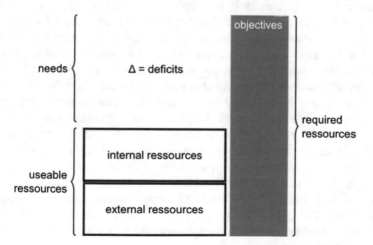

Fig. 8.3 Needs are the difference between necessary and available resources that are relevant for reaching the goal set

are an objectified quality that however always stand in a subject-related context (Hasseler and Görres 2010; Gründger 1977).

This definition deviates from economically inspired concepts of need, which define a need as a requirement that can be met with purchasing power or market demand. For on the one hand, in contrast to goals, individually felt needs can only be argued for from a supra-individual level to a limited extent. On the other hand, as a monetary resource, buying power must be allocated to resources and would therefore be listed twice (Hasseler and Görres 2010; Gründger 1977).

> Example:
> On first consideration, Uli Anders' need can be established through the difference between going to the toilet during the day and at night. The only changing constant is the light. Uli Anders' need lies initially in lighting the way between his bed and the toilet, be it with a torch, a night light or remote control of the regular lighting. However, there is further access to need in the constant environment. Here is the question: why is Uli Anders afraid of falling at night—but he wasn't a year ago? Considering it this way leads to the conclusion that back then Uli Anders would have had the power and speed to get his balance back after colliding with a dresser. Simply moving the furniture or introducing night-time lighting only compensates one resource loss. Uli Anders' actual need is to build up strength and speed, either through movement training at home or collectively in public.

In the BMBF project QuartrBack, the strong focus on the goal of preventing emergencies and providing rapid help when there were emergencies led to an overemphasis of this need situation. This is understandable as the goal of 'security' is attractive for all participants. If 'security' is brought into connection with the risk of disorientation, then the questions arises if the objective should be prevention, compensation or rehabilitation. Emergency is designated as rehabilitative: what is done when someone loses their orientation? This is a question of individual and general resources. Involving the police or rescue services is every bit as imaginable as phoning someone who knows the area and can suggest a direction to go looking. Using a navigation device or human accompaniment or restricting the person to known paths and routes can be designated compensatory. Prevention means shaping processes and structures in such a way that orientation is constantly updated. This can be done with cards, signs and way markings at the structural level and with close comparison of the person's own position with physical and mental cards at the procedural level. By considering the situation in this way, the broad spectrum of needs becomes apparent. The interplay of emergency and daily life also becomes clear. Thus, good signage can prevent loss of orientation, but also compensate and rehabilitate. When it comes to orientation, it can serve both the affected person and third parties.

Tip:
   Differentiate between needs groups and target groups. Needs groups sharpen the focus and answer the question: 'what does a social circle require?' They are narrowly defined groups of people who have the same need, that is to say they exhibit the same resource profile in pursuit of the same objective. Target groups by contrast answer the question: 'which social circle is interested in a given matter?' They widen the focus and address more broadly defined groups, which exhibit a high or increased probability of belonging to a needs group.

8.2.6   Resource Balance Sheet

With the term positive resource balance sheet and the demand for a high degree of individualisation and the retention of everyday skills and promotion of development potential, Lindenberger et al. (2011) introduce three criteria that age-appropriate technologies must satisfy. These criteria address the intervention and evaluation levels equally. The positive resource balance sheet means measures are only justified when the outlay is less than the benefit. If learning results or loss of competence are included, both short and long-term effects must be considered. Furthermore, the objective (created by a third party) and the subjective resource balance sheet my vary. A higher level of individualisation is reached when a measure is in a position (for example, in the interests of a positive balance sheet) to support and claim available resources based on individual standards. The demand to maintain everyday skills and to promote development potential aims to prevent the person being stressed by swings between being overstrained and not challenged enough. When people are overstrained their available resources are subsumed by demands and the burdens of everyday life. This can lead to resignation. When people are not challenged enough, by contrast, too few demands can be put on resources, which can lead to degeneration because of missed training (Lindenberger et al. 2011).

Example:
   For night-time lighting Uli Anders can arrange a torch, a permanent night light, automatic lighting or a remote control. The criteria are (1) cost, (2) influence on sleep and (3) manageability. A torch is cheap and doesn't affect sleep but must be carried to the toilet. A permanent night light is cheap and easy to use, but does disturb sleep with its light emissions. Automatic lighting and remotely controlled lighting come with higher costs, but don't disturb sleep and are easy to use. In the resource balance sheet, automation of the lighting performs poorly—it means that Uli Anders gives up

responsibility for controlling the lighting and its physical implementation. Therefore, Uli Anders decides on a remote control solution. To promote movement, he decides to go to a sports group and do morning exercises in bed. The sports groups brings him particular added value in the area of social participation.

In counselling situations and projects, irritations repeatedly arise when different care goals compete with each other. Thus, in the MOVEMENZ project, a situation was discovered where a person was passively driven to an activity although she could have walked there herself. This contrast was initially interpreted by those on the outside as a failure by the carers. That is understandable, as the focus of the MOVEMENZ project is on remaining mobile and promoting movement. The irritation disappeared when a further dimension of need in the field of participation and activation was opened up. The person affected can't take part in the singing group if she is exhausted. Independent walking would promote movement but endanger participation and activation. Therefore, in this case, the resource balance sheet is only positive if the person affected is passively conveyed to the singing group. The movement can be make up in connection with this activity and at other times of day.

Tip:
    Balance the outlay and benefit of measures from the perspective of all stakeholders. Strengthen common consensus decisions for measures that all participants can experience with a positive resource balance sheet.

8.3   Success Factors for Developing and Implementing Technology

Technology has an influence on the understanding of care through the risk of alienation and greater outlay due to administering the technology. In medically oriented intensive care, the necessary technology leads to an altered human image, for a functional, somatic-technical fundamental understanding is a prerequisite for connecting machines to people. As a consequence, measures concentrate on bio-logical malfunction. Functionality is the most important thing. Deviations from the norm are met with certain standardised actions, care planning is performance and illness oriented, care is judgmentally classified as basic treatment care, and success is measured by the removal of the malfunction (Friesacher 2005). The influence of technology can spread to the whole care arrangements and is a particular factor that co-determines the character of professional and informal care and can also change it. Technology cannot be considered in isolation. It is interwoven with social acts and relationships. Technology, social acts and relationships mutually influence each other in part and beyond that the roles for social actions, the actors and care arrangements are also assigned. Technology can split various activities and assign various groups of people or transfer them 'online' in the virtual space. The resultant changes for the whole arrangement should be taken into consideration (Krings et al. 2014). The consequences of technology for (professional) care activities are wide-ranging and potentially concern all parts of the care process. Against this background, we must handle the development and implementation of technology responsibly.

Quality criteria in care science
The effects and range of technology in care are known through care instruments: instruments provide structure. They are part of the care process. They work clearly and explicitly. However, instruments can depict the care situation in a way that simplifies and reduces it. Whether an instrument is useful or not depends not just on the quality of the instrument but also on the skills of the user (Bartholomeyczik 2009).

Therefore, technology as an instrument must be viewed like other instruments in care and satisfy the demands from a care science perspective that are placed on instruments. For 'instruments that don't lead to subsequent decisions and measures are superfluous in treatment' (Bartholomeyczik 2009). Thus technology either governs actions or is redundant. If it governs actions it can influence matters as profound as the human image and our understanding of care. Therefore, it must be professionally considered and evidence-based.

> Tip:
> Precisely plan the implementation of technology in advance. Make care structures, processes and goals transparent for all stakeholders. Demand transparency in relation to demands on the structures, processes and goals of affected stakeholders—in particular in the development and administration of technology.

8.3.1 Ethical Reflections

Added to the (1) complexity of care arrangements and the (2) limited range of each instrument is the further challenge (3) of people's obedient behaviour (Lüttke 2003). Therefore when using technology, as when using all instruments, 'departure from the rule' must be promoted and demanded in justified individual cases. For this, a strong ethical culture is necessary and corresponding instruments, such as ethical guidelines, ethical case reviews or ethical councils. Professional and ethical reflection must thereby encompass the development process and implementation, as well as regular/permanent use of the technology. As reflection can only happen ex ante (for example, after becoming aware of a question) (Grunwald 1999), in the interests of sustainable development or implementation work, a fundamental sensitisation to ethical questions is useful, and ethical reflection should take place as soon as possible. Sensitisation could be thought of in terms of the categorical imperative: 'Act as though the maxim of your action were by your will to become a universal law of nature.'(Kant, Critique of Practical Reason). This can be consulted neither for moral legitimation nor adequate ethical reflection. This third formulation of the categorical imperative can, however, contribute to uncovering first questions. Here the questions might be: 'What concerns have the consequences of an innovation, development or implementation for myself and/or all of society?' or 'What conditions and exceptions must pertain for me and/or the whole of society for innovation, development and implementation to become law?' (Gethmann and Sander 1999) Moral assessment depends on socially rooted morality and is therefore fundamentally revisable (Grunwald 1999). Against this background, ethical reflection should be used as guidance, so that moral barriers and dangers can be recognised and target values reached. In case of conflicting values, ethical reflection provides the opportunity to protect endangered values (Grunwald 1999). Affected parties should be involved in this reflection, not least for sustainability, particularly when decisions involve palpable imbalances between beneficiaries and risk bearers (Renn 1999). In such as discursive process, it should also be disclosed which human image and which values and moral arguments are governing the development and implementation (Riedel 2015). If concerns exist when disclosing these points that can be a further sign that ethical reflection is necessary. In practice, this manifests repeatedly in products that want to skim off the financial potential of the so-called silver generation or target efficiency improvements with offering all those affected a balanced cost-benefit equation.

Smart technology is increasingly moving into the area of support options for older people, for one thing to support users themselves in the areas of security and comfort, and for another to relieve the heavy physical everyday burden on carers. With this use of technology the question of its moral tenability comes up again. Particularly in situations where different moral values collide, the moral question occurs of what is the right conduct (Ammicht Quinn et al. 2015). Ethics enable us to reflect on what benefit individual people and society derive from technology (Velasquez et al. 1987). In the next decade, the direction of development of our society and social attitudes to older people, will also depend on how we shape the introduction of technology and what expectations we have in connection with it. Technical assistance systems are not just a small practical support in the domestic environment of individual users, they demand a corresponding infrastructure coupled with high investment and a social picture of how we want to assure the care of older people in the future (Manzeschke et al. 2013).

Experience from development history shows that every benefit that technology brings has a side effect. There are currently no clear, comprehensive answers to ethical questions (Körtner 2016). This makes it all the more important to take account of informational self-determination for older people. Users must be included in technical innovations from the outset. They are the metronomes and decide on the construction and possible applications of the system. Technology represents an opportunity: we are in a process that we are accompanying and can help to form—or perhaps must help to form. Older people who need support in their everyday lives invest hope in the technical developments and cherish the desire to live a secure, self-determined 'good life' by using technology. Ancient philosophers around Aristotle were already focused on the 'good life'. Sen and Nussbaum define it in 'The Capability Approach' as a life that allows the maintenance and development of capabilities (Ammicht Quinn et al. 2015). Starting with the thought that the use of technology enables the development and maintenance of capabilities, the question arises of whether it is ethically acceptable to withhold supportive technologies from potential users. To answer this question in detail, the specific benefit of a technical assistance system must be presented in detail. The MEESTAR model (model for the ethical evaluation of socio-technical arrangements) can be used for structured ethical reflection.

Practice examples
In the context of the BMBF-sponsored project, Patronus, an ethical workshop was realised with the help of the MEESTAR model. This contributed to creating a sensitivity to ethical challenges that occur in the course of a project related to technical assistance systems. The interdisciplinary composition of the work groups meant that the ethical questions could be discussed from different perspectives. Furthermore, this ensured that sensitivity to moral unease in the different disciplines was disseminated.

In the course of the workshop, it was shown the moral assessment of individual questions depends heavily on the kind of technology used and must be regularly individually validated. Therefore, every moral problem area that occurs in the context of technological development must be considered from different perspectives and oriented to the individual needs of the technology user. Following the MEESTAR model, the focus should be on the dimensions of welfare, self-determination, security, justice, privacy, participation and self-understanding (Manzeschke et al. 2013). Welfare themes such as giving up capabilities were, for example, considered alarming in the workshop ('People who use technology can be tempted to give up personal capabilities and rely completely on the technology. If, for example, the technology reminds them to take their medication every day, the user can come to rely too much on this function and then doesn't think about it anymore themselves'). Security themes also gave pause for thought, such as the question of data protection ('Depending on the kind of data transferred, corresponding data security must be in place. The more data are transferred, the harder they are to secure').

In general in the implementation of technical assistance systems, the focus must always be on the most significant thing: supporting the user. It's also important to ensure that the implementation of technology doesn't obscure other good solutions.

Tip:

Ethical reflection is occasionally confused with moral concerns and experienced as a barrier. Use ethical reflection as a navigational instrument to circumvent moral barriers—plan ahead to involve them.

8.3.2 Sustainability

When looking at the sustainability of solutions it makes sense to compare the effect and outlay of a measure with alternative measures, against the background of collective targets. Effects include intended as well as unwanted effects. These can be differentiated into direct and indirect. Particularly with measures that cause efficiency enhancements, rebound effects can be observed: for occasionally an objective pre-existing improvement leads to an actual deterioration, because people's behaviour has changed (Buhl and Acosta 2015; Santarius 2012; Pirgmaier and Gruber 2012). To handle rebound effects, measures should not be thought of in isolation (Speck 2016). Faster transport connections don't lead to time savings if longer distances are travelled. Energy efficient components in computers and smartphones don't lead to energy savings if the potential won is deployed in better performance. Better performance can lead to new goals or forms of use and thereby to higher use. Safety measures (helmets, belts, air bags, ABS, occupational safety, assistance systems, condoms, nutrition, reduced tobacco consumption) can lead to riskier behaviour because of strong feelings of security (risk compensation). That means that other people's riskier actions must be reckoned with too (Bioly 2010; Osorio et al. 2015; Jetzek 2009). Included in outlays, after the Total Cost of Ownership (TCO), are outlays for creation, implementation, administration, maintenance, operation and disposal (Gronwald 2015). Particularly in institutional settings, implementation and administration are important for satisfactory application: if rules are implemented that go beyond pure operation/application and refer to individuals or user groups, then carers are affected at the conduct level. Regardless of whether their job is in administering the technology or in representation/ advocacy for the person requiring care, the carers must deal with the technology at a structural level. They must make sweeping decisions and assessments as to whether a individual should/may/can/must be assigned to a particular group or rule. This has an influence on the actual conduct of carers and the allocation of their professional capacities. This influence can persist in the diagnostic view, because for all those involved individual and above all situational conduct is then only possible in the context of the control structure. A diagnostic view outside this context has no consequence and so loses importance. Therefore professional reflection is necessary about how far and in what direction technical administration influences daily care conduct. For this, it must be explicit which obligations and conduct freedoms are affected in daily life and in particular in the care process. Where routines develop, and above all where they are introduced in standardised and automated ways, professional and ethical reflection or evaluation is indicated. Beyond that, on the organisation's side, the range of rules must be considered and a corresponding way created that doesn't just allow the validity of a rule for a particular situation to be based on a professional or ethical justification but requires it. At this point, professional or ethical councils or case conferences can be supportive.

Tip:
Ensure transparency early on concerning outlays for administration and implementation as well as unwanted effects. Put particular emphasis on communication when transferring a solution from the protection of the laboratory into the context of reality.

8.3.3 Security and Risk

What is understood by security is the result of a social negotiation process. Security consists of a triad of components: value level, potential threat and protection level. The value level describes an asset worth protecting as well as its significance. This area is subjective, meaning that security is not clearly defined, being described, for example, as 'a situation in which a citizen is protected from threats—either by neutralisation, avoidance or low-risk behaviour' (Ziegleder et al. 2011). Security is a term that depends on the values and value systems of individual citizens. That's one reason why absolute security is not attainable (Müller 2015; Ziegleder et al. 2011; Daase and Deitelhoff 2013). If values compete in one person or among several people, ethical reflection can be used as an evaluation method. Threat potential combines threats, the likelihood of them happening and the potential scale of damage with risk. A clear evaluation of risk parameters must often be conditional. Establishing probability and damage scales helps to make threat potential more concrete in the face of unclear data (Müller 2015; Regenfus and Vieweg 2010).

Protection level subsumes protection measures and their weighting. They usually reduce the likelihood of a threat happening or its effects (Müller 2015).

The BMBF project QuartrBack tried to protect the protection-worthy assets of movement and participation in social life from the threat potential that wandering brings with it. The level of protection should be defined according to the individual threat potential in a needs-based approach. Therefore, to develop proactive measures the general threat potential must be established before determining the individual threat potential. However, currently there are no reliable figures on the frequency (likelihood of a threat happening) and affects (scale of damage) of running away and wandering among people with dementia. The police helicopter squadron in Baden-Württemberg provides information on the frequency of missing person searches (generally, not specifically for people with dementia): in 2015 there were 942 searches, and in 2014 there were 949. The police and Federal Border Guard and the 'Stelle zur trägerübergreifenden Qualitätssicherung im Rettungsdienst Baden-Württemberg' (Baden-Württemberg rescue service) cannot provide statistics on key words such 'helpless person'. Under the search term 'dementia', the 'blue light' section of the press portals lists: (http://www.presseportal.de/blaulicht/suche. htx?q=demenz) on 20.02.2016, 188 results from 2004 to 2016, of which 49 are reports from 2014 and 58 reports from 2015. Overall, there were 70 missing person reports and 25 people were found again. The portal reports on three deaths, seven spontaneous instances of help being offered without a preceding missing person report and seven fraudulent cases. The figures are not valid, but they do justify asking the question of whether there is a need for action at the state or federal level. To put the figures in the context of the overall population: in 2014 the population of Baden-Württemberg was 10.7 million, of whom 2.1 million were aged 65 and over and 1.1 million were aged 75 and over (https://www.statistik.baden-wuerttemberg. de/BevoelkGebiet/Bevoelkerung/99025010.tab?R=LA from 1.7.2016).

Poor data can neither make an efficient contribution to developing measures nor help with people's sense of security. Relatives and carers suffer repeatedly from the justified fear that a person with dementia might not find their way home. Sense of security describes security as absence of fear or anxiety from certain dangers. It is a multidimensional structure that exhibits an affective, a cognitive and a conative dimension (Ziegleder et al. 2011; Daase and Deitelhoff 2013).

Therefore, in QuartrBack defined scenarios, technologies and processes are developed with a describable and therefore manageable risk. Precise protection measures can be derived for these narrowly defined risks. Outside these areas, a risk culture must be established in everyday life. This involves weighing up values: does the person with dementia's freedom of movement prevail or their welfare with the need avoid potential harm? At this point it's less about a contextual decision for one value or another. It's much more about being aware that there is a conflict of values. It's about highlighting that one value has been decided upon and justified after careful consideration.

> Tip:
>   The need for security is based on ascription of values. The term security is a nebulous social convention. Make the term concrete by talking about the risks that a measure ensure against.

8.3.4 *Transparency and Information Flow*

In the BMBF project *Patronus an assistance syst*em was developed that automatically recognised challenging situations and autonomously launched a solution. Amongst other things, it encompassed emergency detection and help initiation and it was adjusted to the life and health situation of individual users and to their wishes and ideas. Underlying the project was Baltes' selective optimisation and compensation (Baltes et al. 1998), which describes capabilities in older age. By developing remaining resources and capabilities as far as possible, maximum quality of life is targeted and resource-based action strategies derived. At the beginning of the project a needs analysis was conducted to collect demands for the system from the point of view of older people and service providers. In the context of the assistance model the general needs of elderly people and general offerings in various areas were captured in a data bank, and the system was tested in a practical trial. The practical trial showed that the use of complex technologies requires a high degree of coordination. Furthermore, cooperation with third parties, not people primarily involved in the project (for example, electricians) played a deciding role in whether installation was successful. Because of the complexity of technical assistance systems, it became clear that a project leader with technical understanding needed to be in situ when the supportive technological solution was installed at home to bring the individual technical components together and coordinate the operation. Responsibilities had to be clearly defined and set out in a binding installation timetable.

From a social service point of view, the delivery of information and explanations to the older users plays an important role. The advantages and disadvantages of a technical system must be made transparent, so that each individual user can decide what added value the system's installation brings for them. Depending on their technical biographies, older people tend to react more sluggishly to new technological developments than younger people, because they must devote more time to learning new technologies (Lamsfuß 2012). For this reason, a large number of senior citizens initially consented to take part in the trial of the Patronus system because they had a lot of faith in the social institutions and their employees. This underlines the importance of enthusing employees about project proposals and the associated technical installation. Employees who are sceptical about a system convey that scepticism to the user consciously or unconsciously. Alongside reliable support in the test surroundings, previous experiences with technology installation play an important role. Residents who had already had bad experiences with technology installation in their homes were rather sceptical about the system. Comprehensive explanations of the system, for example in the context of information events, continual communication with the residents and corresponding transparency concerning the functionality of the system could lead to the users successfully completing the live test. Here, transparency on purchase costs and how they would be split was also important.

Tip:
Clarify the following questions and themes:

- *Organisation installation*: How complex will the system installation be? Define responsibilities. Establish a project manager. Get an installation timetable from the project manager.
- *Communication with residents*: Provide transparency. Clarify all essential themes. Cover things that are supposedly obvious. When is the system in active mode? Who reacts how to what (for example, system failure with emergency call triggering)? Who is the contact person for which areas?
- *Information materials*: Leave information materials in the user's home. Draw their attention in particular to contact people. Appeal to visitors as well in the information materials and explain to them any rules of conduct or restrictions on informational self-determination.
- *Costs*: Clarify early on what the costs are and who will pay them. These include, for example, conversion measures, energy supply and maintenance. Is the purchase price part of the investment costs or is it possible to get the care insurance to take on these costs?
- *Test function*: Make the system accessible by, for example, integrating a test mode like those found in smoke detectors. It's important for the residents to be able to reassure themselves that the system is working at any time—especially if it's a security system designed to be as unobtrusive as possible.

8.4 Conclusion

Thanks to the experience of this project, recognition is growing that technical as well as non-technical innovations should be founded on professional and ethical reflection as well as critical and open communication. Take your time. Speak to the people involved. Look for the small solutions that make a difference in everyday life—then you will automatically stumble upon the big challenges.

References

Ammicht Quinn R, Beimborn M, Kadi S et al. (2015) Alter. Technik. Ethik. Ein Fragen- und Kriterienkatalog. Eberhard Karls Universität Tübingen, Tübingen
Andrews J (2012a) How acute care managers can support patients with dementia. J Nurs Manag 19 (2):18–20. doi:10.7748/nm2012.05.19.2.18.c9062

Andrews J (2012b) A nurse manager's guide to support patients with dementia. Int J Older People Nurs 24(6):18–20

Baillie L, Merritt J, Cox J (2012) Caring for older people with dementia in hospital. Part one: challenges. Int J Older People Nurs 24(8):33–37

Baltes P, Lindenberger U, Staudinger U (1998) Life-span theory in developmental psychology. In: Lerner R (ed) Handbook of child psychology. Theoretical models of human development. Wiley, New York, pp 1029–1143

Bartholomeyczik S (2009) Standardisierte Assessment instrumente: Verwendungsmöglichkeiten und Grenzen. In: Bartholomeyczik S, Halek M (eds) Assessmentinstrumente in der Pflege. Möglichkeiten und Grenzen, 2nd edn. Schlüttersche Verlagsgesellschaft, Hannover, pp 12–26

Bartholomeyczik S, König P, Boldt C et al. (2006) Entwicklung und Anwendung der ICF aus der Perspektive der Pflege. Positionspapier der deutschsprachigen Arbeitsgruppe ICF und Pflege. Available via DIALOG. http://www.deutsche-rentenversicherung.de/cae/servlet/contentblob/206966/publicationFile/2309/icf_dokumentation_4_anwenderkonferenz.pdf. Accessed 27 Jul 2016

Bioly S (2010) Demografischer Wandel, Decarbonisierung und steigende Verkehrsleistung. In: Klummpp M (ed) Dienstleistungsmanagement in Theorie und Praxis, vol 13. Logos, Berlin

Buhl J, Acosta J (2015) Work less, do less? Working time reductions and rebound effects. Sustain Sci 11(2):261–276

Daase C, Deitelhoff N (2013) Privatisierung der Sicherheit. Forschungsforum Öffentliche Sicherheit, Berlin

Dörner D (1987) Problemlösen als Informationsverarbeitung. Kohlhammer, Stuttgart

Duncker K (1966) Zur Psychologie des produktiven Denkens. Reprint der Originalausgabe von 1935. Springer-Verlag, Berlin

Elsbernd A (2013) Konzepte für die Pflegepraxis: Theoretische Einführung in die Konzeptentwicklung pflegerischer Arbeit. In: Einführung von ethischen Fallbesprechungen. Ein Konzept für die Pflegepraxis. Ethisch begründetes Handeln praktizieren, vol 3. Jacobs Verlag, Lage, pp 13–38

Friesacher H (2005) Pflegeverständnis. In: Ullrich L, Stolecki D, Grünewald M (eds) Intensivpflege und Anästhesie. Thieme, Stuttgart, pp 38–45

Gethmann CF, Sander T (1999) Rechtfertigungsdiskurse. In: Grunwald A, Saupe S (eds) Ethik in der Technikgestaltung. Praktische Relevanz und Legitimation. Schriftenreihe der Europäischen Akademie zur Erforschung von Folgen wissenschaftlich-technischer Entwicklungen. Wissenschaftsethik und Technikfolgenbeurteilung, vol 2. Springer-Verlag, Berlin, Heidelberg, pp 117–151

Gronwald KD (2015) Integrierte Business-Informationssysteme. ERP, SCM, CRM, BI, Big Data Analytics. Prozesssimulation, Rollenspiel, Serious Gaming. Springer-Verlag, Berlin, Heidelberg

Gründger F (1977) Zum Problem der Bedarfsermittlung bei Investitionen im Bildungs- und Gesundheitswesen. Eine Vergleichende Untersuchung unter besonderer Berücksichtigung des Schul- und Krankenhaussektors. Volkswirtschaftliche Schriften 255. Duncker und Humbolt, Berlin

Grunwald A (1999) Ethische Grenzen der Technik? Reflexionen zum Verhältnis von Ethik und Praxis. In: Grunwald A, Saupe S (eds) Ethik in der Technikgestaltung. Praktische Relevanz und Legitimation. Schriftenreihe der Europäischen Akademie zur Erforschung von Folgen wissenschaftlich-technischer Entwicklungen. Wissenschaftsethik und Technikfolgenbeurteilung, vol 2. Springer-Verlag, Berlin, Heidelberg, pp 221–252

Hasseler M, Görres S (2010) Was Pflegebedürftige wirklich brauchen. Zukünftige Herausforderungen an eine bedarfsgerechte ambulante und stationäre pflegerische Versorgung, 2nd edn. Schlütersche Verlagsgesellschaft, Hannover

Hölzle C (2011) Gegenstand und Funktion von Biografiearbeit im Kontext Sozialer Arbeit. In: Hölzle CJ, Jansen I (eds) Ressourcenorientierte Biografiearbeit. Grundlagen. Zielgruppen. Kreative Methoden, 2nd edn. VS Verlag für Sozialwissenschaften, Wiesbaden, pp 31–54

Hülsken-Giesler M (2015) Technik und Neue Technologien in der Pflege. In: Brandenburg H, Dorschner S (eds) Pflegewissenschaft 1. Hogrefe, Bern, pp 262–276

Illich I (1998) Selbstbegrenzung. Eine politische Kritik der Technik. C. H. Beck, München

Institut für europäische Gesundheits- und Sozialwirtschaft Berlin (2014) Unterstützung Pflegebedürftiger durch technische Assistenzsysteme. Available via DIALOG. http://www. heilberufe-online.de/kongress/rueckblick/berlin2014/Braeseke-Grit_Unterstuetzung-Pflegebeduerftiger-durch-technische-Assistenzsysteme.pdf. Accessed 27 Jul 2016

Jetzek F (2009) Conjoint- und Discrete-Choice-Analyse als Präferenzmessmodelle zur Beurteilung des präventivmedizinischen Risikoverhaltens. Universität Passau, Passau, Theorie und computergestützte Umsetzung unter Verwendung von SMRT

Körtner T (2016) Ethical challenges in the use of social service robots for elderly people. Z Gerontol Geriatr 49(4):303–307. doi:10.1007/s00391-016-1066-5

Koutoukidis G, Strainton K, Hughson J (2013) Tabbner's Nursing Care. Theory and Practice, Elsevier, Chatswood

Krings BJ, Böhle K, Decker M et al. (2014) Serviceroboter in Pflegearrangements. In: Decker M, Fleischer TH, Schippl J et al. (eds) Zukünftige Themen der Innovations- und Technikanalyse. Lessons learned und ausgewählte Ergebnisse. KIT Scientific Reports 7668. KIT Scientific Publishing, Karlsruhe, pp 63–122

Lamsfuß R (2012) Nur kein Schnickschnack. Eine soziologische Betrachtung der Internetnutzung in der Generation 50plus. In: Kampmann B, Keller B, Knippelmeyer M et al (eds) Die Alten und das Netz: Angebote und Nutzung jenseits des Jugendkults. Gabler Verlag, Springer Fachmedien, Wiesbaden, pp 12–26

Lindenberger U, Lövdén M, Schellenbach M et al. (2011) Psychologische Kriterien für erfolgreiche Alterstechnologien aus Sicht der Lebensspannenkognition. In: Lindenberger U, Nehmer J, Steinhagen-Thiessen E et al. (eds) Altern in Deutschland, vol 6. Nova Acta Leopoldina N. F., Vol 104 (368) Verlagsgesellschaft, Stuttgart, pp 17–33. Available via DIALOG. https://www. demografie-portal.de/SharedDocs/Downloads/DE/Studien/AiD_Altern_Technik.pdf?__blob= publicationFile&v=2. Accessed 31. Jul 2016

Lüer G, Spada H (1992) Denken und Problemlösen. In: Spada Hans (ed) Lehrbuch Allgemeine Psychologie, 2nd edn. Hans Huber, Bern, pp 189–280

Lüttke HB (2003) Gehorsam und Gewissen. Verlag Peter Lang, Frankfurt am Main, Die moralische Handlungskompetenz des Menschen aus Sicht des Milgram-Experimentes

Mahler R (2012) Resilienz und Risiko. Ressourcenaktivierung und Ressourcenförderung in der stationären Suchttherapie. In: Petzold HG, Lammel A, Leitner A (eds) Integrative Modelle in Psychotherapie, Supervision und Beratung, Springer VS, Wiesbaden

Manzeschke A (2013) Technik in der Altenhilfe—Fluch oder Segen? BruderhausDiakonie. Sozial. Magazin für Politik, Kirche und Gesellschaft in Baden-Württemberg 2013(2):16

Manzeschke A, Weber K, Rother E et al (2013) Ergebnisse der Studie "Ethische Fragen im Bereich Altersgerechter Assistenzsysteme". VDI/ VDE Innovation + Technik GmbH, Berlin

Müller KR (2015) Handbuch Unternehmenssicherheit. Umfassendes Sicherheits-, Kontinuitäts- und Risikomanagement mit System, 3rd edn. Springer, Wiesbaden

Osorio A, Lopez-del Burgo C, Ruiz-Canela M et al (2015) Safe-sex belief and sexual risk behaviours among adolescents from three developing countries: a cross-sectional study. BMJ Open 5(5):1–9. doi:10.1136/bmjopen-2015-007826

Pirgmaier E, Gruber J (2012) Zukunftsdossier Alternative Wirtschafts- und Gesellschaftskonzepte. Reihe "Zukunftsdossiers" N3. Bundesministerium für Land- und Forstwirtschaft, Umwelt und Wasserwirtschaft (Lebensministerium), Wien

Regenfus T, Vieweg K (2010) Sicherheit- und Risikoterminologie im Spannungsfeld von Technik und Recht. In: Winzer P, Schnieder E, Bach FW (eds) Sicherheitsforschung. Chancen und Perspektiven, Springer, Heidelberg, pp 131–144

Renn O (1999) Ethische Anforderungen an den Diskurs. In: Grunwald A, Saupe S (eds) Ethik in der Technikgestaltung. Praktische Relevanz und Legitimation. Schriftenreihe der Europäischen Akademie zur Erforschung von Folgen wissenschaftlich-technischer Entwicklungen. Wissenschaftsethik und Technikfolgenbeurteilung, vol 2. Springer, Berlin, pp 63–94

Riedel A (2015) Ethikberatung in der Altenpflege. Forderungen und Gegenstand. In: Coors M, Simon A, Stiemerling M (eds) Ethikberatung in Pflege und ambulanter Versorgung. Modelle und theoretische Grundlagen. Jacobs Verlag, Lage, p 45–67

Röhner J, Schütz A (2016) Psychologie der Kommunikation. In: Kriz J (ed) Basiswissen Psychologie. Springer-Verlag, Wiesbaden

Santarius T (2012) Der Rebound-Effekt. Über die unerwünschten Folgen der erwünschten Energieeffizienz. Impulse zur Wachstumswende, Wuppertal Institut für Klima, Umwelt, Energie GmbH, Wuppertal

Schiemann D, Moers M (2004) Werkstattbericht über ein Forschungsprojekt zur Weiterentwicklung der Methode "Stationsgebundene Qualitätsentwicklung in der Pflege". Available via DIALOG. https://www.dnqp.de/fileadmin/HSOS/Homepages/DNQP/Dateien/Weitere/WerkstattberichtSQE.pdf. Accessed 31 Jul 2016

Schuntermann M (2006) Die internationale Klassifikation der Funktionalität, Behinderung und Gesundheit (ICF) der Weltgesundheitsorganisation (WHO). Kurzeinführung. Available via DIALOG. http://www.deutsche-rentenversicherung.de/cae/servlet/contentblob/206970/publicationFile/2307/icf_kurzeinfuehrung.pdf. Accessed 31 Jul 2016

Senge PM (2011) Die fünfte Disziplin. Kunst und Praxis der lernenden Organisation. 11th edn. Schäffer-Poeschel, Stuttgart

Speck M (2016) Konsum und Suffizienz. Eine empirische Untersuchung privater Haushalte in Deutschland, Springer, Wiesbaden

Thieme (2015) I care—Pflege. Thieme, Stuttgart

Velasquez M, Andre C, Shanks TSJ et al (1987) What is ethics? Issues Ethics 1(1):1–2

Weinberger N, Decker M (2015) Technische Unterstützung für Menschen mit Demenz? TATUP - Technikfolgenabschätzung Theorie und Praxis 24:36–45. Available via DIALOG. http://www.tatup-journal.de/downloads/2015/tatup152_wede15a.pdf. Accessed 31 Jul 2016

Weiß C, Lutze M, Compagna D et al. (2013) Abschlussbericht zur Studie "Unterstützung Pflegebedürftiger durch technische Assistenzsysteme". BMG, Berlin. Available via DIALOG. https://www.vdivde-it.de/publikationen/studien/unterstuetzung-pflegebeducrftiger-durch-technische-assistenzsystemc/at_download/pdf. Accessed 31 Jul 2016

WHO (2002) Towards a common language for functioning, disability and health. ICF. The international classification of functioning, disability and health. Available via DIALOG. http://www.who.int/classifications/icf/icfbeginnersguide.pdf. Accessed 31 Jul 2016

Ziegleder D, Kudlacek D, Fischer TA (2011) Zur Wahrnehmung und Definition von Sicherheit durch die Bevölkerung Erkenntnisse und Konsequenzen aus der kriminologisch-sozialwissenschaftlichen Forschung. Forschungsforum Öffentliche Sicherheit, Berlin

Zwingmann E (2007) Über das gemeinsame (Be-)Finden: von Ressourcen, Lösungen und Wohl-Befinden. In: Frank R (ed) Therapieziel Wohlbefinden. Ressourcen aktivieren in der Psychotherapie, Springer, Heidelberg, pp 103–115

Chapter 9
Parents' Experiences of Caring for a Ventilator-Dependent Child: A Review of the Literature

Minna Lahtinen and Katja Joronen

Abstract The purpose of this review was to describe parents' experiences of caring for a ventilator-dependent child at home. The data consisted of 19 articles searched from Medic, Arto, Linda, Cinahl, Medline and PsycINFO databases and manually for the 1998–2016. The data were analysed using qualitative content analysis. Parents' experience of caring for a ventilator-dependent child at home showed that caring for such a child made life a constant battle for survival. This constant battle for survival meant struggling with life management challenges, trying to maintain balance within the family and turning to everyday resources. The home became like a hospital because of the presence of technology and healthcare staff. Parenting of a ventilator-dependent child meant undertaking some nursing. The effects of this on parenting have not been given enough attention in paediatric care. The results of the review show that the life with a ventilator-dependent child is a constant struggle and balancing act, which the parents cannot manage without professional support. Further research on the topic is needed because more and more increasingly sick children are able to live at home with the help of medical technology.

Keywords Children · Parents · Review · Respiration artificial · Ventilator-dependent

9.1 Background

Medical technology has developed rapidly since the 1990s. Most pre-term and seriously sick and injured children survive. This increased survival rate has resulted in a growing population of families caring for children with complex needs at home

M. Lahtinen
Tampere University Hospital, Science Center and School of Health Sciences, University of Tampere, (FI), Tampere, Finland

K. Joronen (✉)
School of Health Sciences, University of Tampere, (FI), Tampere, Finland
e-mail: katja.joronen@uta.fi

I. Kollak (ed.), *Safe at Home with Assistive Technology*,
DOI 10.1007/978-3-319-42890-1_9

(Elias and Murphy 2012). In the US, this small but rapidly growing group is referred to as technology-dependent children (Montagnino and Mauricio 2004; Earle et al. 2006). These children need medical devices to compensate for malfunction caused by childhood diseases or injury. Medical technology allows children to survive and prevents further disabilities (Office of Technology Assessment 1987). According to the Office of Technology Assessment, four groups of children can be identified whose reliance on medical devices and nursing care to maintain life mean they can be classified as technology-dependent:

1. children dependent for at least part of each day on mechanical ventilators
2. children requiring prolonged intravenous delivery of nutritional substances or drugs
3. children dependent on a daily basis on other device-based respiratory or nutritional support, including tracheotomy tube care, suctioning, oxygen support and tube feeding
4. children with prolonged dependence on other medical devices that compensate for vital body functions and who require daily or near-daily nursing care.

The latter group includes (Office of Technology Assessment 1987, p. 17):

- infants requiring apnea (cardiorespiratory) monitors
- children requiring renal dialysis as a consequence of chronic kidney failure
- children requiring other medical devices such as urinary catheters or colostomy bags, as well as substantial nursing care, in connection with their disabilities.

In this review, the target group is group 1 of this technology-dependent population: children dependent for at least part of each day on mechanical ventilators. Ventilatory support to maintain adequate gas exchange and oxygenation in children may be initiated because of a wide variety of often rare medical diagnoses, including bronchopulmonary dysplasia, central hypoventilation syndrome, degenerative neuromuscular diseases, congenital anomalies and spinal cord injuries. Furthermore, ventilatory support may be required either continuously (24 h a day) or partially (for example, during sleep) (Earle et al. 2006; Vuori and Ylitalo-Liukkonen 2009).

In the US, the number of children receiving home mechanical ventilation (HMV) increased from eight in the period 1979–1988 to 122 in the period 1999–2008. This increase was most obvious in the youngest age group with invasive HMV (Paulides et al. 2012). Most countries, including Finland, do not have a national register of children who use HMV. Research into parents' experiences of caring for children at home already exists, and it is known that parenting a chronically ill child is demanding. However, recent literature reviews have not been reported (Lillrank 1998; Wang and Barnard 2002; Boling 2005; Hopia 2006). This chapter gathers previous research literature about parents' experiences of caring for ventilator-dependent children.

9.2   The Purpose of the Review

The purpose of this chapter is to describe parents' experiences of children's ventilator treatment based on previous studies for the years 1998–2016. In reviewing the literature, the research question was: what are the experiences of parents who care for ventilator-dependent children at home?

9.3   Materials and Methods

The data search process is presented in Fig. 9.1. Inclusion criteria were: parents who have experience of caring for a ventilator-dependent child aged 0 to 18 years old. Database limits were that the child should be aged zero to 18 years, the paper should be in English or Finnish, should be peer-reviewed and should have been published between 1998 and 2012. The search was updated in July 2016. The primary data consisted of 18 articles searched systematically from Medic, Arto, Linda, Cinahl, Medline and PsycINFO databases and manually for the years 1998–2012. The search was updated for the years of 2013–2016 in June 2016, and one new research article was included. Thus, the final data set was 19 articles. The data were analysed by qualitative content analysis.

9.4   Description and Evaluation of the Quality of the Material

All the selected research articles (n = 19) were peer-reviewed research papers. The quality of the studies from the first search (n = 18) was assessed and scored based on the background and purpose of the research paper, materials and methods, reliability and ethics, as well as the results and conclusions. This quality assessment scale was modified from Turjamaa's et al. (2011) checklist and the scale used ranged from one to three (1 = poor, 2 = moderate, 3 = good). Average total scores were calculated for every reviewed paper. The average scores ranged from 2.2 to 2.8, and the maximum value was 3.0. Thus, the reviewed articles were mainly high quality. However, ethical issues were discussed in the studies in less depth.

From the selected studies 13 papers were qualitative (1, 2, 3, 4, 5, 6, 7, 14, 15, 16, 17, 18 & 19), and two were quantitative studies (9 and 11). Four studies included both quantitative and qualitative research methods (8, 10, 12 and 13). Six of the studies were done in the US (8, 9, 11, 14, 15 and 17), five in Canada (1, 2, 3, 12 and 18), five in the UK (5, 6, 7, 10 and 13), one in Australia (16), one in Norway (4) and one in Sweden (19).

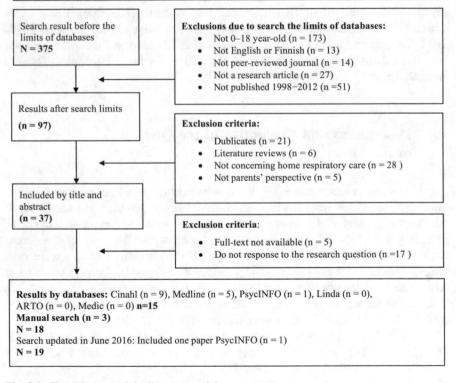

Research question:
What are the experiences of parents who care for their ventilator-dependent child at home?

Search terms:
"Ventilators, Mechanical"(PsycInfo: mechanical ventilation*) , Ventilator Patients", Home respiratory care, Respiration, Artificial (PsycInfo:Artificial respiration) ,"technology dependent","Parents", "Family"(PsycInfo:Family Members") tai "Caregivers" & "Attitude" , experience, view tai perspect / hengityslaite, hengityskone, hengityshalvaus, lääkintälaite tai respiraattori & lapsi

Databases (1998-2012):
Cinahl (n = 160), Medline (n = 175), PsycINFO (n = 18),
Linda (n = 3), ARTO (n = 0), Medic (n = 19)

Inclusion criteria:

- Concerns 0–18 year-old ventilator-dependent child
- Parents' perspective
- Responses to the research question

Search result before the limits of databases
N = 375

Exclusions due to search the limits of databases:
- Not 0–18 year-old (n = 173)
- Not English or Finnish (n = 13)
- Not peer-reviewed journal (n = 14)
- Not a research article (n = 27)
- Not published 1998–2012 (n =51)

Results after search limits
(n = 97)

Exclusion criteria:
- Dublicates (n = 21)
- Literature reviews (n = 6)
- Not concerning home respiratory care (n = 28)
- Not parents' perspective (n = 5)

Included by title and abstract
(n = 37)

Exclusion criteria:
- Full-text not available (n = 5)
- Do not response to the research question (n =17)

Results by databases: Cinahl (n = 9), Medline (n = 5), PsycINFO (n = 1), Linda (n = 0),
ARTO (n = 0), Medic (n = 0) **n=15**
Manual search (n = 3)
N = 18
Search updated in June 2016: Included one paper PsycINFO (n = 1)
N = 19

Fig. 9.1 Flow diagram of the literature search

9.5   Data Analysis

The reviewed studies were analysed by using inductive content analysis (Kylmä and Juvakka 2007). The studies were recorded on a table, with the author, year of publication, country, the purpose of the research, data and research methods, main results, and average scores from the quality appraisal.

The data were first read carefully and expressions of parents' experiences were extracted from them. To summarise the information, the primary expressions were cut down and similar expressions were grouped into 16 subcategories. Thirdly, subcategories were abstracted into three categories through conceptualisation. Finally, the three categories were abstracted to one main category, which responded to the research questions. All the categories were named according to their content.

9.6   Results

Parents' experience of caring for a ventilator-dependent child at home showed that caring for such a child made life a constant battle for survival. This constant battle for survival meant struggling with life management challenges, trying to maintain balance within the family and turning to everyday resources.

The home became like a hospital because of the presence of technology and healthcare staff. Parenting of a ventilator-dependent child meant undertaking some nursing. The effects of this on parenting have not been given enough attention in paediatric care.

9.6.1   Struggling with Life Management Challenges

Parents' struggle with life management challenges included the challenge of coping with everyday management, the unpredictability of life, physical exhaustion, mental distress and isolation.

The challenge of everyday management referred to parents' experience of life's complexity (1, 3, 5 and 14). Parents had to learn new skills (17 and 19) in order to be able to take care of their child whose care needs at home were demanding. Furthermore, the implementation of home healthcare was often very challenging (3 and 19). Some parents found the medical devices complicated and frightening (16). The challenge of everyday management was increased by the substantial amount of time required for child care at home (5, 14, 15 and 17). Caring for a child at home weakened the family's financial situation because the costs of caring for the child at home were high (3, 5, 6, 10, 14, 17 and 18) and parents' employment opportunities were scant because of time spent in child care (5, 6 and 10).

Parents found the unpredictability of life exhausting (6 and 14). Life with a ventilator-dependent child was a constant process of adapting to change (14) and dealing with uncertainty (3, 5 and 14). The child's state of health required parents to be constantly vigilant (6, 14 and 19) and prepared for emergency situations (6 and 19). It was also difficult for parents to make even short-term plans, to say nothing of long-term future plans (3, 5 and 14).

Home care of a ventilator-dependent child was physically exhausting (3, 6, 14, 16 and 18). Parents' sleep at night was often disturbed due to the complex needs of the child (3, 5, 6, 9, 11, 17, 18 and 19). Child care was also very tying and dominated life (6 and 19). Parents had very little free time (3, 5, 6, 9 and 18) and they thought that there was too little home care assistances available (3, 5, 6, 7, 10, 11 and 13).

Parents' mental distress included feelings of being overwhelmed by life (1, 3, 5 and 16). Parents were frustrated (13, 14 and 16), stressed (3, 6, 9 and 16), tense (2, 3 and 6) and depressed (8), and they suffered feelings of inadequacy (14). The constant fear of losing the child was emotionally difficult (3 and 14). Parents' experience of life's injustice and the lack of help available with child care sometimes made them feel angry and bitter (13).

Parents reported feeling a deep sense of isolation (2, 3, 8, 14, 16 and 18), and they felt confined to their home (3, 8 and 16). Taking the child outside the home was difficult (3, 5, 6, 16 and 18). Social contacts outside the home were rare (5, 6, 14, 16, 18 and 19) and maintaining relationships was difficult (3, 5, 8, 14 and 16). On the one hand, difficulties finding qualified home care assistance (5, 6, 7, 11 and 13) meant social relationships diminished further, and many parents were very lonely (2). On the other hand, the frequent presence of health care personnel at home made it difficult to maintain social relations because it was difficult to invite guests to the house (7 and 14). Some parents felt like outsiders (2), stigmatised (2 and 6) and excluded (2, 16 and 18) from society.

9.6.2 Maintaining Balance Within the Family

Parents trying to maintain balance within the family referred to balancing the child's and the family's needs, multi-dimensional parenting, a changed relationship with their spouse, a hospitalised home environment and conflicts between the family and health care organisations (Fig. 9.2).

Parents found balancing the child's and the family's needs very challenging, because the ventilator-dependent child's care needs took time from other family members (4, 5, 14, 15 and 19). Siblings got less attention, which made the parents feel guilty (3 and 6). Parents had no time for themselves because they prioritised the needs of other family members over their own needs (4, 14 and 18). Primary responsibility for child care often fell on mothers (5). Parents strove to maintain normal family routines but it was difficult (2, 14, 15, 18 and 19). Parents also worried about the future of the family (14). The presence of home nursing staff

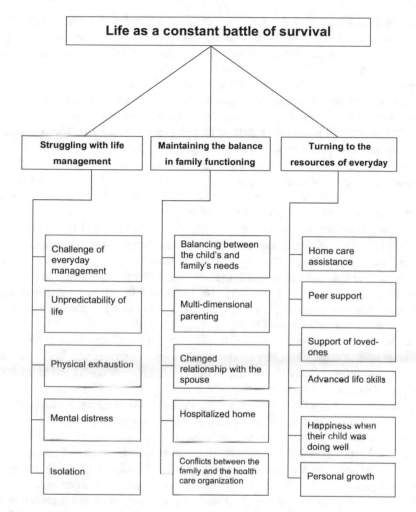

Fig. 9.2 Parents' experience of caring for ventilator-dependent child at home

affected family relations by reducing explicit expressions of emotion between family members (7).

Multi-dimensional parenting meant that it addition to their roles as mother or father, parents had to play the role of nurse (2, 7, 16, 17 and 19) and lawyer (2, 4, 14, 17 and 19). The role of nurse to their own child affected them strongly emotionally. In particular, performing unpleasant treatments on their child was distressing, stressful and broke the parenting experience (7 and 17). Parents were also afraid that some treatments might harm their child (7). They constantly struggled with how to be a good parent (3 and 7). Some parents said they constantly sought information and learned to rely on their own knowledge about their child's needs (19). Rearing a ventilator-dependent child created several challenges. On the one hand, parents wanted to raise their sick child normally; on the other hand setting boundaries was more difficult than usual and parents found they were overprotective. Parents wanted more guidance and advice on rearing a child with special needs (15 and 18). They also thought that their parenthood was sometimes threatened by social institutions and their rules and regulations (19).

The home care of a ventilator-dependent child greatly changed the parents relationship. Continuous lack of time alone impaired the relationship (3, 5, 6, 7 and 14). Parents wanted to spend more time face to face (5, 6 and 14). The presence of home health care professionals reduced displays of affection between parents; the presence of night nurses in particular affected the intimacy of the parental relationship (7). Although caring for a child at home affected relationships negatively in many ways and in some cases led to divorce (14), some of the parents said shared hard experiences had strengthened their relationship (5).

Parents' experience of a hospitalised home environment was the result of the continual presence of medical equipment and home carers and professionals in their home. Parents found that technology dominated their home (6, 7, 14 and 16), which transformed it into a mini-hospital (7). Although home care staff were seen as essential in terms of coping (1, 5, 6, 11, 16, 17, 18 and 19), these strangers changed the home to a hospital-like (7, 16, 19) and public (16 and 19) place. Parents thus suffered from lack of privacy (7, 10, 15, 16, 17 and 19).

Conflicts between the family and health care organisations made it difficult to maintain normal family life. The availability of support services was inadequate and accessible services often did not respond to the family's needs (4, 6, 13 and 19). Some parents said that some assistants did not respect their family values, leading to a transgression of private boundaries (19). Furthermore, battling bureaucracy was difficult, and this could result in insurance claims and even lawsuits (19). Home carers were also changed often, and the continuity of care was flawed (4 and 19). Parents and professionals had different views about care policies and procedures (7 and 19). Parents wanted closer monitoring of child care and better setting of treatment goals (13 and 19). They were disappointed because their expertise gained from experience was not valued (6, 7 and 19), and they were not included in the making of treatment decisions (4 and 6). In addition, parents felt that the emotional impact (of the home treatment) on parents was not taken into account (6 and 10).

9.6.3   Turning to the Everyday Resources

Parents turned to everyday resources, which were home care assistance, peer support, the support of loved ones, advanced life skills, happiness when their child was doing well and personal growth (Fig. 9.2).

Home care assistance helped parents to cope (1, 5, 6, 10, 11, 13, 15, 16, 17, 18 and 19). Home care assistance made life more flexible (6 and 19), gave the parents a breather (16) and better sleep (5 and 11), as well as helping them to manage everyday life (6, 18 and 19). Home care staff helped parents to learn how to care for their child at home (17), and they supported parents' mental well-being (6 and 19). Some parents described how caring relationships with care assistants, based on reciprocal trust, brought friendship, energy and joy to the family (19). However, home care assistance was not available widely enough (3, 5, 6, 7, 10, 11 and 13). Delivery of care supplies to the home was a big help for families with a ventilator-dependent child (5).

Parents found peer support essential (12, 14, 17 and 19). Peer support helped them to cope (12 and 17), made them feel understood (12 and 19), gave them a sense of togetherness with others (19), and it helped them to learn how to treat their child (12 and 17). Peer support created and maintained hope (17), although some parents said it was difficult to take if the child of a peer family had better prospects (12).

The support of loved ones was of great significance to parents. The support of loved ones helped them to cope and reduced stress, feelings of isolation and symptoms of depression (8, 9, 14 and 17). In particular, the importance of grand-parents was emphasised as a source of practical and emotional support (14, 17 and 18). At the same time, the parents occasionally felt guilty about taking a lot of help from grandparents (14). Maintaining a close partnership with their spouse was a challenge (3, 5, 6, 7 and 14), but this relationship was an important resource for parents (17). Other children in the family helped in everyday life (17 and 19) and parents found friends to be an important source of support (9 and 17).

Parents mentioned advanced life skills as being important everyday resources. Organisational skills were necessary to manage everyday life, and these skills were developed through experience (1, 14 and 17). Creativity, using their skills, flexibility and humour helped parents to cope (14). Parents' medical knowledge coupled with their expertise gained from experience enabled them to provide individual treatment (7 and 19). Routines maintained order and helped with the pacing of everyday life (17 and 19). Some parents described how they tried to create a safe place of their own in the presence of professionals (19).

The child's well-being was important to the parents. The child itself was a great resource for the parents and their love for their child helped them to cope (3, 7 and 19). Daily life included many moments of joy (1 and 3) and parents experienced caring for their child as rewarding and worth the effort (3). Parents described how their children had a great will to live and the child's friends made the parents less important (19). Even small advances in the child's health created hope (1 and 3). Parents were happy that it was possible to care for their child at home (4), and they found that medical devices helped them to monitor the child's condition and improved their quality of life (5).

Parents reported personal growth as human beings. They found that their coping skills were improved and they were proud that they'd survived (4 and 16). Personal growth included a wider perspective on life (17), a new hierarchy of values (14), and endurance and resilience (19). Some parents found religion to be an important resource (17). The child's illness affected their world view and highlighted the positive aspects of life (14). Parents tended to adapt to and accept things they couldn't able change and to change things they were able to influence (4, 14 and 19). Facing facts and remaining strong enabled growth (17). Some parents described maintaining a positive attitude to life (4 and 16) and they found that their ability to empathise and face difficulties had improved (16).

9.7   Discussion

The literature review results showed that, according to how parents describe it, life with a ventilator-dependent child is a constant battle for survival. This constant battle for survival means struggling with life management challenges, trying to maintain balance within the family and turning to everyday resources. Whiting (2014) suggests that the sense that the parents of children with complex health needs make of their situation plays a crucial role in determining how they experience the impact of the disability and the need for support.

Life management challenges included, for example, the unpredictability of life and physical exhaustion. The results are consistent with previous research results concerning chronically ill children generally. It has been noted that the various practical problems in the somatic treatment of chronically ill children and in managing their everyday life exhausted parents and impaired their quality of life (Boling 2005; Hopia 2006). In addition, difficulties in tolerating the uncertainty and planning for the future were found in earlier studies concerning parents of somatically ill children (Lillrank 1998; Hopia 2006). One reason for physical exhaustion is that parents are sleep deprived. A recent systematic review shows that the sleep deprivation experienced by parents of children with complex needs is both relentless and draining. This sleep deprivation affects the parents' ability to function. Sleep disturbance also has the potential to change family relationships, in some cases causing marital strain (McCann et al. 2015).

Parents of ventilator-dependent children found their lives on the one hand to be emotionally distressing and isolated, and on the other hand bright and rewarding, which enabled personal growth. Child care took a lot of time, and the child became the centre of family activities. These challenges were also reported in previous studies examining parents' experience of dealing with seriously ill children (Boling 2005; Hopia 2006). A previous review by Lindahl and Lindblad (2013) also suggested that families with very ill children are at risk being excluded from society and everyday relationships with other people. The results of a previous review (Wang and Barnard 2002) on parents' experiences of caring for a ventilator-dependent child were similar to this review, recording psychological distress and challenges in balancing family life. It is noteworthy that the previous review did not explain the positive effects of home care on parents and family (Wang and Barnard 2002). Only a few studies have reported positive changes in parents and families with a chronically ill child. One exception is the study by Wong and Chan (2005); whose results indicated that an ill child could reinforce family cohesion. Furthermore, according to Hopia's study (2006), parents experienced a child's illness in part as positive because dealing with difficult issues brought family members closer together.

Parenting a technology-dependent child involved nursing, which parents saw as emotionally burdensome. Wang and Barnard (2002) also indicated that medical interventions performed by parents had a negative affect on the parenting experience. The unpleasant affect of technology was emphasised in parents' experiences of caring for a child at home. Home became a mini-hospital and the loss of privacy was considered oppressive. These results confirm the findings of the literature review by Wang and Barnard (2002), Jachimiec et al. (2015).

This review showed that one of the concrete resources in everyday life was a home care service, which was seen as essential for coping with everyday life. In addition, parents perceived their improved life skills as a significant resource. Also peer support and support from loved ones helped parents to survive; in particular, they emphasised the importance of feeling understood. Mikkonen (2009) has suggested that peer support helps people to feel that they are not alone with their life situation and problems. The starting point of nursing interventions should always be a family's unique situation and their particular need for assistance.

While life with a technology-dependent child is a constant struggle for survival, parents do want to take care of their child at home. Earlier international research results suggest that treatment at home is also more beneficial to society. However,, the negative economic affects fall more on families (Office of Technology Assessment 1987; Jacobs und McDermott 1989; Wang and Barnard 2002).

9.8   Conclusions

1. A technology-dependent child at home completely changes family life. Sufficient home care services are essential for parents to cope and survive.
2. Parenting of a technology-dependent child includes the role of nurse; the emotional effects of this have not been taken into account enough in patient education and in assessing parents' need for support.
3. In addition to social support and improved life skills, parents of a technology-dependent child's main resources are the satisfaction they get from the child's well-being and their personal growth.
4. Parents think that home is the best place for a sick child, even though this means the home is transformed into a mini-hospital due to medical technology and health care personnel being in the home. Loss of privacy at home upsets parents.

9.9 Recommendations for Nursing Practice

When a child is seriously ill, parents' coping skills are put to the test. It is essential that enough support is organised for families. Every mother and father has the right to be primarily a parent and not a nurse to their own children. Healthcare professionals should support mothers and fathers in parenting. However, it should be acknowledged that parents are the decision-makers in their child's care (Mendes 2013).

To help the family to survive and ensure safety, parents must be well versed in the specific requirements of their child's care. Too little attention has been paid to the experiences of family carers who manage complex technical health procedures at home (McDonald et al. 2016). In addition, it is important to arrange home care assistance, particularly at night.

It is more economical for society for children's long-term ventilator care to be carried out at home than in hospital. However, society must provide medical equipment and meet the cost of home care professionals. The health care provided must include adequate overnight respite for families (McConkey et al. 2011). There is no question that being in the home environment with their parents benefits children's growth and development.

References of the Review

1. Alexander E, Rennick JE, Carnevale F et al (2002) Daily struggles: living with long-term childhood technology dependence. Can J Nurs Res 34(4):7–14
2. Carnevale FA (2007) Revisiting Goffman's Stigma: the social experience of families with children requiring mechanical ventilation at home. J Child Health Care 11(1):7–18. doi:10.1177/1367493507073057
3. Carnevale FA, Alexander E, Davis M et al (2006) Daily living with distress and enrichment: the moral experience of families with ventilator-assisted children at home. Pediatrics 117(1):48–60
4. Dybwik K, Tollali T, Nielsen EW et al (2011) "Fighting the system": families caring for ventilator-dependent children and adults with complex health care needs at home. BMC Health Serv Res 11:156 − 164. doi:10.1186/1472-6963-11-156
5. Heaton J, Noyes J, Sloper P et al (2005) Families' experiences of caring for technology-dependent children: a temporal perspective. Health Soc Care Community 13(5):441–450. doi:10.1111/j.1365-2524.2005.00571.x
6. Kirk S, Glendinning C (2004) Developing services to support parents caring for a technology-dependent child at home. Child Care Health Dev 30(3):209–218. doi:10.1111/j.1365-2214.2004.00393.x
7. Kirk S, Glendinning C, Callery P (2005) Parent or nurse? The experience of being the parent of a technology-dependent child. J Adv Nurs 51(5):456–464. doi:10.1111/j.1365-2648.2005.03522.x

8. Kuster PA, Badr LK (2006) Mental health of mothers caring for ventilator-assisted children at home. Issues Ment Health Nurs 27(8):817–835. doi:10.1080/01612840600840588

9. Kuster PA, Badr LK, Chang BL et al (2004) Factors influencing health promoting activities of mothers caring for ventilator-assisted children. J Pediatr Nurs 19(4):276–287. doi:10.1016/j.pedn.2004.05.009

10. Margolan H, Fraser J, Lenton S (2004) Parental experience of services when their child requires long-term ventilation. Implications for commissioning and providing services. Child Care Health Dev 30(3):257–264. doi:10.1111/j.1365-2214.2004.00414.x

11. Meltzer LJ, Boroughs DS, Downes JJ (2010) The relationship between home nursing coverage, sleep, and daytime functioning in parents of ventilator-assisted children. J Pediatr Nurs 25(4):250–257. doi:10.1016/j.pedn.2009.01.007

12. Nicholas DB, Keilty K (2007) An evaluation of dyadic peer support for caregiving parents of children with chronic lung disease requiring technology assistance. Soc Work Health Care 44(3):245–59. doi:10.1300/J010v44n03_08

13. Noyes J, Hartmann H, Samuels M et al (1999) The experiences and views of parents who care for ventilator-dependent children. J Clin Nurs 8(4):440–450

14. O'Brien ME (2001) Living in a house of cards: family experiences with long-term childhood technology dependence. J Pediatr Nurs 16(1):13–22. doi:10.1053/jpdn.2001.20548

15. O'Brien ME, Wegner CB (2002) Rearing the child who is technology dependent: perceptions of parents and home care nurses. J Spec Pediatr Nurs 7(1):7–15

16. Wang KK, Barnard A (2008) Caregivers' experiences at home with a ventilator-dependent child. Qual Health Res 18(4):501–508. doi:10.1177/1049732307306185

17. Wilson S, Morse JM, Penrod J (1998) Absolute involvement: the experience of mothers of ventilator-dependent children. Health Soc Care Community 6(4):224–233

18. Woodgate RL, Edwards M, Ripat J (2012) How families of children with complex care needs participate in everyday life. Soc Sci Med 75(10):1912–1920. doi:10.1016/j.socscimed.2012.07.037

19. Lindahl B, Lindblad BM (2013) Being the parent of a ventilator-assisted child: perceptions of the family-health care provider relationship when care is offered in the family home. J Fam Nurs 19(4):489–508. doi:10.1177/1074840713506786

References

Boling W (2005) The health of chronically ill children: lessons learned from assessing family caregiver quality of life. Fam Community Health 28(2):176–183

Earle RJ, Rennick JE, Carnevale FA et al (2006) "It's okay, it helps me to breathe": the experience of home ventilation from a child's perspective. J Child Health Care 10(4):270–282. doi:10.1177/1367493506067868

Elias ER, Murphy NA (2012) Home care of children and youth with complex health care needs and technology dependencies. Pediatrics 129(5):996–1005

Hopia H (2006) Somaattisesti pitkäaikaissairaan lapsen perheen terveyden edistäminen. Akateeminen väitöskirja, Acta Universitatis Tamperensis 1151. Tampere University Press, Tampere

Jachimiec JA, Obrecht J, Kavanaugh K (2015) Interactions between parents of technology-dependent children and providers: an integrative review. Home Healthc Now 33 (3):155–166. doi:10.1097/NHH.0000000000000205

Jacobs P, McDermott S (1989) Family caregivers costs of chronically ill and handicapped children: method and literature review. Public Health Rep 104(2):158–163

Kylmä J, Juvakka T (2007) Laadullinen terveystutkimus. Edita, Helsinki

Lillrank A (1998) Living one day at the time. Parental dilemmas of managing the experience and the care of childhood cancer. Stakes tutkimusjulkaisuja 89. Gummerrus, Jyväskylä

Lindahl B, Lindblad BM (2013) Being the parent of a ventilator-assisted child: perceptions of the family-health care provider relationship when care is offered in the family home. J Fam Nurs 19(4):489–508. doi:10.1177/1074840713506786

McCann D, Bull R, Winzenberg T (2015) Sleep deprivation in parents caring for children with complex needs at home: a mixed methods systematic review. J Fam Nurs 21(1):86–118. doi:10.1177/1074840714562026

McConkey R, Gent C, Scowcroft E (2011) Critical features of short break and community support services to families and disabled young people whose behavior is severely challenging. J Intellect Disabil 15(4):252–268. doi:10.1177/1744629511433257

McDonald J, McKinlay E, Keeling S et al (2016) Becoming an expert carer: the process of family carers learning to manage technical health procedures at home. J Adv Nurs 72(9):2173–2184. doi:10.1111/jan.12984

Mendes MA (2013) 'Parents' descriptions of ideal home nursing care for their technology-dependent children. Pediatr Nurs 39(2):91–96

Mikkonen I (2009) Sairastuneen vertaistuki. Akateeminen väitöskirja. Acta Universitatis Kuopioensis 173. Kuopion yliopisto, Kuopio

Montagnino B, Mauricio R (2004) The child with a tracheostomy and gastrostomy: parental stress and coping in the home—a pilot study. Pediatr Nurs 30(5):373–380

Office of Technology Assessment (1987) Technology-dependent children: Homecare versus hospitalcare, a technical memorandum. US government, Washington DC

Paulides FM, Plötz FB, Verweij-van den Oudenrijn LP et al (2012) Thirty years of home mechanical ventilation in children: escalating need for pediatric intensive care beds. Intensive Care Med 38(5):847–852. doi:10.1007/s00134-012-2545-9

Turjamaa R, Hartikainen S, Pietilä A-M (2011) Kotona asuvien iäkkäiden ihmisten voimavarat—systemoitu kirjallisuuskatsaus. Tutkiva Hoitotyö 9(4):4–13

Vuori A, Ylitalo-Liukkonen K (2009) Vaikean neuromuskulaarisen hengitysvajepotilaan hoidon järjestäminen Varsinais-Suomen sairaanhoitopiirissä. Available via DIALOG. http://www.docstoc.com/docs/106917745/Vaikean-neuromuskulaarisen-hengitysvajepotilaan-hoitoprosessi-VSSHPssa-ver-090929-LOPULL. Accessed 1 Jan 2013

Wang KW, Barnard A (2002) Technology-dependent children and their families: a review. J Adv Nurs 45(1):36–46

Whiting M (2014) What it means to be the parent of a child with a disability or complex health need. Nurs Child Young People 26(5):26–29. doi:10.7748/ncyp.26.5.26.e390

Wong M, Chan S (2005) The coping experience of Chinese parents of children diagnosed with cancer. J Clin Nurs 14(5):648–649. doi:10.1111/j.1365-2702.2004.01009.x

Chapter 10
Evaluation and Outcomes of Assistive Technologies in an Outpatient Setting: A Technical-Nursing Science Approach

Ulrike Lindwedel-Reime, Alexander Bejan, Beatrix Kirchhofer and Peter Koenig

Abstract As our society is ageing, the ability to live a self-determined life is growing more and more important for an ever increasing number of elderly people. Assistive technologies (AT) for outpatient use created by interdisciplinary research and development teams can be helpful assets in that regard. However, assistive systems or processes have to be engineered and evaluated with care in order to finally qualify as evidence-based products that will in turn be accepted by their intended users as well as the market. For that purpose, the involved researchers and developers coming from different areas of expertise—with potentially divergent focuses—need to be aware of and have an in-depth knowledge regarding assessment instruments that can be used to empirically evaluate the outcomes of AT as thoroughly as possible. In order to provide an overview of the relevant subject matter, this chapter initially introduces the reader into the basic nature of the concept "assistive technology" and the general idea of evaluation. In the next step, different models of the development and evaluation process relating to AT are examined from a technical perspective followed by an analysis of the evaluation process from the healthcare perspective. Ultimately, a synthesis of technology and evaluation from both angles of vision is proposed as a holistic approach and enriched by a number of scientifically tested assessment instruments usable for outpatient AT evaluation.

Keywords Assistive technologies (AT) · Research and development (R&D) · Evaluation · Assessment instruments · Nursing

U. Lindwedel-Reime (✉) · A. Bejan · B. Kirchhofer · P. Koenig
Faculty Health, Safety, Society, Furtwangen University, Furtwangen, Germany
e-mail: liru@hs-furtwangen.de

© The Author(s) 2017
I. Kollak (ed.), *Safe at Home with Assistive Technology*,
DOI 10.1007/978-3-319-42890-1_10

10.1   Introduction

To be able to live a largely self-determined life is a great asset for numerous senior citizens. In order to facilitate this for as many people as possible, particularly in the context of an ageing society, researchers and developers in the interdisciplinary field of human-technology interaction are trying to develop innovative assistive technologies (AT). While already offering many possibilities to develop and market helpful products, this field is still a challenging area associated with considerable research and development efforts. Despite many attempts to produce marketable and functional products, only a few have actually been implemented into practice. Their use in everyday care—at least in Germany—is correspondingly low.

In this regard, various barriers to innovation have been identified in the meantime. Beside the aspects of user orientation and fitness for the intended purpose, these include the legal framework and requirements for training and further education. Molenbroek (2013) argues that the benefit of the technical assistance systems that have been developed is usually not apparent to the end user, cannot be conveyed to them and often isn't explained. This is more than surprising, especially when you consider that communicating and explaining technology to older people provides high added value—particularly in relation to acceptance which in the final analysis forms the basis for an implementation that makes economic sense (Molenbroek 2013).

Another problem is that the various influencing factors affected by the implementation of such a technology are not sufficiently incorporated, if at all, in the development and testing, either because they haven't been clearly defined or because they are simply ignored. Often only the technical aspects are evaluated, such as usability or acceptance by users. For a long time, systematic reviews on the topic have criticised the poor nature of the evidence on the actual effects of AT (Connell et al. 2008; Topo 2009; Martin et al. 2008; Anttila et al. 2012). Aspects such as particiaption and quality of life as well as dimensions that are also relevant in care, for example the carers' work flow, should be included more in evaluations. There is also increasing demand and support for this in the care research agenda (Behrens et al. 2012). Basically, there still seems to be a lack of experience in implementation and evaluation and a lack of validated instruments in the area of human-technology interaction and AT to support people who need assistance.

This chapter presents a structured overview of the processes and instruments used to evaluate technical aids as well as problems with implementing them in practice, particularly in the outpatient environment, so that carers as well as technicians may see "beyond their own noses" and become familiar with current instruments for assessment and also become aware of how to put them to good use. Starting with diverse definitions of terms and classifications for AT and for their evaluation, a synthesis is generated through a short summary of the development and testing processes as seen from the technical and from the care perspective. This is rounded off at the end by listing a range of established assessment instruments. By deliberately taking a cross-domain approach and evaluating individual AT products in an appropriate way, useful and usable systems can be invented and developed that potentially benefit all.

10.2   Classification of Assistive Technologies

In Germany, the English term 'assistive technologies' is often considered to be synonymous with German terms for rehabilitation technology and support technology. What is meant by that is all technical aids—often in the context of aftercare following an unsuccessful treatment or instead of a treatment that isn't currently available—that remedy or at least partially compensate for compromised bodily functions or performance, restricted participation and barriers in a person's surroundings. Specific technical aids for rehabilitation should ideally be adjusted to suit individual functional and social limitations by multi-disciplinary medical, technical and care science development teams. By contrast, the focus of the German term 'medical technology', which is better known in the healthcare sector, is rather prevention, diagnosis and therapy with an aim of curing patients rather than compensating for lost capacities (Zagler 2013).

As early as 1990 AT devices were defined in the US Individuals with Disabilities Education Act (IDEA) as objects, devices and systems either commercially purchased, modified or adjusted and used to maintain or improve the capabilities of people with disabilities (IDEA 1990). In theory at least, Americans with disabilities were henceforth to be given the same educational opportunities as people without disabilities through a legal right to access appropriate assistive aids or individual educational programmes. Under this definition, everyday low-tech devices such as special pens, magnifying glasses and special feeding bowls count as AT, but above all considerably more complex and powerful high-tech devices such as electric wheelchairs and alternative computer input devices, such as computer-based systems to support communication, help with reading and writing, and controlling surroundings (esp. regarding ill and disabled people) and everyday life. Such technologies don't always have to be invented from scratch, but can also be created by modifying existing—originally not assistive—objects. In Europe, the standard EN ISO9999 'aids for people with disabilities—classification and terminology', currently in its 2011 version, can be taken as a reference. Under this classification, AT can be sorted into different groups of aids, which in turn can divided into two further levels. Thus, a talker system would be grouped under ISO 22 21 12 (ISO level 1:22 aid for communication and information; ISO level 2:21 aid for local communication; ISO level 3:12 software for local communication).

Well-defined classifications can be useful in various respects: being aware of which ISO category (and thus which setting) an AT that is to be developed or investigated falls into allows appropriate instruments to be chosen and used for meaningful testing. It also makes obvious that AT should primarily be researched, developed, implemented and marketed for people with disabilities and not for the sake of the technology itself. This is an important realisation that allows evaluation to be extended to include missing human and contextual dimensions.

10.3 Evaluating Assistive Technologies

10.3.1 Definition of the Term 'Evaluation'

In recent years, evaluation has gained more and more importance in science (Stockmann and Meyer 2014). However, the definition of evaluation varies greatly by context. While Nutbeam and Bauman (2006) see evaluation as a purely formal process for assessing worth, others interpret the term in a considerably broader way (Nutbeam and Bauman 2006). To get the greatest benefit from an evaluation, the following basic questions should be considered: *what* (object) is being evaluated and *why* (to what end), using *which criteria* by *whom* and *how* (methods)? Thus evaluation is understood as a specific action, whose objective is to generate empirical knowledge on the basis of which valuations can be undertaken. Ideally, empirical instruments will be deployed to generate knowledge in this systematic process. Evaluation is also an instrument for quality control (Friedrich et al. 1997).

The principles of evaluation vary in many respects according to whether they are seen from a health care science or a technical perspective. In that sense, it is useful to look at both perspectives separately before linking them to facilitate a considerate, interdisciplinary evaluation that adds value in the context of assistive systems.

10.3.2 Development and Evaluation of AT from a Technical Perspective

To gain a better understanding of the need for a thorough evaluation in the context for AT, it is beneficial first to understand the characteristics of the (scientific) development process and its procedural models. To this end, a short overview of common procedural models, characteristics and neuralgic problem points in the development of AT will be given at this point.

Research and development of AT is basically oriented towards defined basic processes from science (particularly technical disciplines connected with human-technology interaction) and (project and product management) practice. In the first step, hypotheses concerning the starting point and effect of special assistive systems—ideally not yet available on the market—that meet particular needs are generated in an exploratory and pre-emptive way by a multi-disciplinary team helped by diverse, user-centred (if appropriate integrating the target group itself) design and research methods.

In the second step, appropriate technologies based on the above are implemented as prototypes or evaluators, (interim)tested/(interim)evaluated in the field or in the laboratory environment taking into account the target group(s) and in the best case further developed interactively in several cycles (in the case of a sequential operation the prototype will only undergo one development process), before their assistive benefit, usability and effects are themselves evaluated in a final step with the target group.

Agile evaluation and development that generates new information in several short iterations is generally recommended, especially in scientific projects, to enable a quick and flexible reaction to changing circumstances and possible intervention—for example, if an AT prototype isn't accepted by the target group, doesn't fulfil its supportive function or only partially fulfils it, new information comes to light, or the hypothesis (and with it the assistive approach) is shown to be suboptimal early on. The illustrations below contrast a simplified sequential and a simplified agile-iterative model as rough basic processes (Figs. 10.1 and 10.2).

10.3.3 Evaluation from a Health Care Science Perspective

Care science evaluation methods encompass the entire spectrum of research methods, ranging from simple, exploratory studies to complex, randomised studies. Heterogeneous methods mixes are used increasingly often alongside classic 'single' qualitative and quantitative methods. Which method exactly should be used depends on the question, the deployed AT as well as the test subjects, and is usually determined at the start of the investigation or research project.

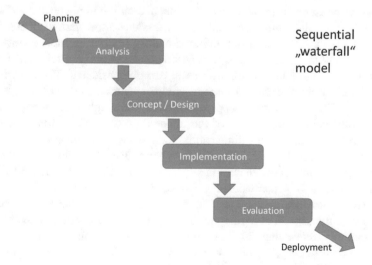

Fig. 10.1 Generalised sequential procedural model (own presentation)

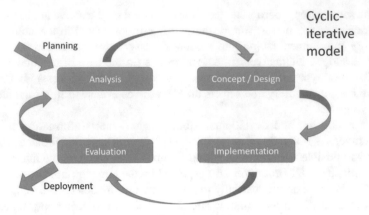

Fig. 10.2 Generalised cyclical-iterative procedural model (own presentation)

It is essential to make this decision—for example, for an application for ethical approval—but in the area of human-technology interaction it does sometimes mean that only a few test scenarios and limited innovation and flexibility are the consequence. From the perspective of technical players, this has the consequence that implemented technology usually lags behind what is possible or is already outdated when it is tested, because established—often inflexible—study protocols are pursued first. In this sense, evaluation from the health care science perspective aims—above all—to research impacts and effects.

To enable measurement of the various effects of implemented AT, validated outcomes and assessment instruments are needed. Depending on the nature of the effect, it's possible to use several instruments. Even in the area of AT, various known assessment instruments are often used in practice. Nonetheless, it is best to define all dimensions that could have an influence on the implemented technology early on. Various definitions are available for this that allow various dimensions to be elicited from a global perspective. Individual factors, for example quality of life or acceptance of the technology, as well as structural factors (e.g. relatives, informal and formal carers and their work processes) can be considered. The influence and effects of AT on the living environment (environmental factors) can then be taken into account. This wide-ranging approach, which sheds light on as many factors as possible influencing the situation of the affected person, is also found in various care theories. By describing dimensions such as what it means to be human, health/illness, care and the environment, an attempt is made to describe networks of relationships and interdependencies in care work. This makes it easier to learn how to deal with AT. The International Classification of Functioning, Disability and Health (ICF) provides another way to capture the various dimensions of AT's effects. Here, the aspects of participation, body function and environmental factors are particularly interesting, as these have not yet been fully researched in an empirical-scientific way, although numerous suitable instruments are available for this (Fig. 10.3).

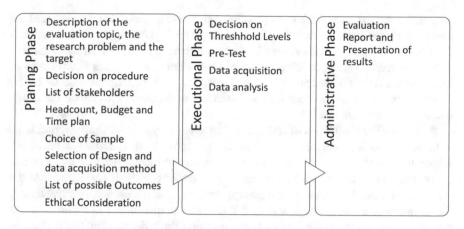

Fig. 10.3 Modified ideal-typical evaluation process based on Stockmann and Meyer (2014)

10.3.4 Synthesis of a Holistic Evaluation Approach

The technical and care perspectives on evaluating technical assistance systems are difficult to unite (Intille 2013). As a result, evaluations are normally carried out from a purely technical design-oriented or from a purely care science perspective. To be able to evaluate AT efficiently and innovatively, coordination well as attention to specifics is needed.

In contrast to industrial technology development projects, whose products are often targeted at one—clearly specified—area and are realised using expert knowledge from that same domain, the evaluation of AT is distinguished by the fact that multi-disciplinary medical, healthcare and technical aspects must be brought together, and for that reason research and development mainly takes place in inter- or trans-disciplinary working groups.

To create a productive (internal) working environment as well as mutual understanding and empathy—in the sense of the 'Design Thinking' of Kelley (Kelley 2007; Brown and Wyatt 2015)—regarding the different working cultures of the care and social science partners and the technical and practical project partners should be sought. In addition to personal openness to the colleague's different areas of expertise, 'meta' documents can be created, for example in the form of collective agreements or glossaries of technical terms that make it easier to see things from the other perspective.

Dealings with (external) players and stakeholders, for example clients or target groups, should also be based on mutual appreciation and consideration. As early as possible in the conception phase, AT should be aligned as closely as possible—assessing the specific needs of the users—to the particular human and social needs of the target group, not least in the context of the human or user-centred design process (abbreviated to HCD/UCD; Norman 2013; ISO 2010), which targets high usability and user acceptance. For people of advanced age, products should take account, amongst other things, of additional sensomotoric deficiencies and be designed in such a way that they can be used by 'technology novices' without frustration. For people with dementia, knowledge of specific individual cognitive deficits or technological facilities that aim to obviate these deficits should help to influence the development process to guarantee the 'user experience' and usability. Evaluations after each product iteration step are indispensable to check the effectiveness of measures used.

Alongside the challenges of internal and external project development, problems with access to test persons can arise in the context of (interim) evaluations. Depending on the time scale of the evaluation requirements, practice partners can prevent access to test persons—e.g. for ethical reasons—which can jeopardise the entire development process. As test groups are usually small (in non-medical areas), the drop-out rate can be high and as AT research is qualitative in nature, serious problems can ensue, which can call into question the evidence for the AT's usefulness and effectiveness.

In the context of AT development using iterative, user-centred design, diverse (interim) evaluation methods can be combined. This can, for example, happen in the form of contextual analysis (contextual design; Beyer and Holtzblatt 1999) where real episodes from practice are observed by the researchers in situ, in the form of taking notes and analysing them, or in the form of scenario-based design (Rosson and Carroll 2003), where fictitious (but based on genuine/typical conditions/facts) personas (prototype users) acting in hypothetical but authentic problem scenarios (prototype use situation) are described and used as a basis for the design.

In the AT research area in particular, target groups are usually diverse and there is no one-size-fits-all solution. Therefore, in most cases, it's sensible and indeed necessary to use an agile-iterative method mix for the conception, development and evaluation of meaningful products.

This is the only way to ensure that qualitatively high-value care and efficient care management on the one hand, and innovative and helpful AT on the other, are united and realised in a meaningful way. To offer a starting point for a holistic evaluation of AT systems, the following section describes assessment instruments whose evaluation dimensions go beyond the customary technical orientation of usability and user acceptance tests and bring in social, environmental and human factors.

10.4 Assessment Instruments to Capture Domestic Dimensions

AT can have an influence on various aspects of a person's life and living environment. Ideally, standardised and successfully tested assessment instruments should be used when analysing these areas. Assessment in this case is the process whereby reliable and relevant information about an individual, a group or an institution is collected in order to make information-based decisions about further actions (Bartholomeyczik and Halek 2010). However, this definition alone naturally reveals nothing about the kind of information collection to be undertaken. The structure of the assessment instruments may differ significantly and depends on the defined aim of the assessment.

The kinds of assessment instruments used and their characteristics vary widely. Alongside standardised structured instruments that estimate an outcome with the help of statistical scores, there exist assessment instruments that are more reminiscent of a guide and that don't calculate scores for estimates. The questions put are more open and variable, which is supposed to enable a wider range of answers and individual views (Bartholomeyczik and Halek 2010). However, the results are harder to interpret and not appropriate for drawing causal conclusions/inferences.

In addition—when it comes to assessment instruments—empirical research quality criteria must be set out. Generally, this involves the three classic criteria of objectivity, validity, and reliability (Reuschenbach 2011; Bortz and Schuster 2010). As these criteria come from classical test theory, which has an empirical-quantitative character and originates in psychology, they are statistically calculated using diverse psychometric values. In the research process, using validated instruments increases the probability that the data obtained is reliable and valid and not based on coincidental results. Insufficient objectivity can lead, amongst other things, to measurement errors. Reliability is often interpreted as dependability and gives, amongst other things, information as to whether an investigation using a particular measuring instrument gives a repeatable result. However, high reliability doesn't necessarily mean that high validity exists. Conversely, there can be no high validity without good reliability, and this in turn is impossible without adequate objectivity. By emphasising the most important criterion—validity—it is possible to check if the measuring instrument is actually measuring the construct that is supposed to be measured (Reuschenbach 2011; Bortz and Schuster 2010).

Various assessment instruments in the area of environmental aspects and care provision are set out below. These should give an idea of which other factors could be included in an evaluation.

10.4.1 Environmental Factors

Environment plays a big role in the treatment of older people in outpatient—but also inpatient—care. The physical and also social environment count as environmental factors. These can be physical aspects such as the condition of the home in terms of barrier-free access or the nature of the facilities, but also factors such as climate and various other beneficial or obstructive factors. Besides family support, the social environment includes other resources, for example, help from neighbours or professional health carers.

In this regard, it should be mentioned that, in the domestic environment in particular, aspects such as a quick and stable Internet connection or the installation and care of AT take on an intrinsic role. It's important to ensure good transfer of information and explanations, particularly if the use of these technologies requires substantial structural changes. This is just as true for conversion work across the whole spectrum of barrier-free installations as it is for unobtrusive things such as the smallest sensors or beacons (i.e. Bluetooth transmission devices).

In rural areas, older Internet connections can be a primary reason why technologies fail. If (wireless) connections exist, they are often not comparable with broadband connections in towns. Alongside malfunctions, unstable connections and dead spots must be taken into account and worked around. Among the target group of elderly people who don't have an Internet connection, it's important to explain clearly what the point of getting one is and also the associated costs. In such a case, personal and monetary resources must be deployed efficiently and in a needs-oriented way.

With regard to this background, the following assessment instruments seem suitable for evaluating technical assistance systems.

1. Neighbourhood-level cohesion and disorder (Cagney et al. 2009): Cohesion and disorder in the local area are determined using 14 questions. The answers are largely recorded using a four-stage Likert scale. Only individual questions are open. The instrument is based on subjective self-assessment. The instrument was tested on residents aged 65 and over in three neighbouring districts on the outskirts of southern Chicago. This assessment instrument first and foremost checks the social environment and in the opinion of the researchers gives good and reliable results concerning the condition of the neighbourhood. An interesting question is whether this instrument would provide a comparable result in a rural environment, as it was originally tested in an urban environment.

2. Usability in my home (UIMH): This assessment instrument was developed to obtain opinion and attitudes from residents about their own domestic environment (Fänge and Iwarsson 1999). This instrument is mainly concerned with the physical environment. It's important to note explicitly here that it's about subjective impressions and that the instrument is based on self-assessment. Residents themselves are to decide how helpful or obstructive their domestic environment is for them. The instrument contains 31 questions on various aspects of the physical domestic environment, which are assessed using a seven-stage Likert scale. The content validity and internal consistence were determined and classified as satisfactory. However, assessment of quality criteria was carried out with an inhomogeneous group of test persons. The group consisted of people across a wide age range from different social groups with diverse functional limitations.

3. Facilitators and barriers survey (FABS): This instrument establishes which environmental conditions facilitate or limit participation in active life for people with limited mobility (Gray et al. 2008). The target group consisted of people with limited mobility particularly in their lower extremities. Elements were

used in the development of this assessment instrument that had been determined beforehand in focus groups. Hence it contained 61 questions in six dimensions. This instrument was also concerned with subjective self-assessment. To start with, elements from people's own domestic environment were ascertained—for example, stairs, access to the bath or which aids were necessary—that could prevent someone from taking part in social life. But the direct surroundings also played a role. Analysis was made, for example, of whether they were wheelchair-friendly and if lifts or windows that open automatically were available in public buildings. In addition, non-structural elements were examined, such as climate or geographic conditions. Finally, social surroundings and their role were integrated into the assessment by asking about support resources. Overall, this instrument delivers very comprehensive and wide-ranging data, though it also requires a big investment of time (Gray et al. 2008).

4. Urban-identity scale: This questionnaire was developed as early as 1988 in the context of town planning (Lalli 1992). The urban identity scale mainly analyses the urban environment in a particular region. The emphasis is more on personal identification with the environment and less on participation options for people with limitations. This is reflected in, for example, questions about employment opportunities in the area and school and shopping options. However, resources for medical treatment and care are also investigated. Overall, the instrument comprises 20 questions in five dimensions, which should be assessed using a five-stage scale. Here again the focus is on subjective self-assessment. However, the questionnaire wasn't scientifically validated and if it were to be used for empirical research this would have to be rectified.

5. Neighbourhood environment walkability scale (NEWS): This instrument is supposed to capture the extent to which the environment facilitates sporting activities and thus it is targeted at the physical environment (Saelens et al. 2003). It's not designed for people with assistance needs, but results from the assessment instrument can also be used for this purpose. The instrument consists of 83 questions, divided into nine dimensions, which are supposed to be answered on a four or five-stage scale using subjective self-assessment. The questionnaire was validated for scientific research purposes by investigating the sporting activities of two different communities using this instrument. The reliability was graded high ($\alpha = 0.63$–0.80), hence the questionnaire is suitable for scientific purposes.

6. Craig Hospital inventory for environmental factors (CHIEF): This instrument captures the barriers and obstacles that might prevent people with assistance needs from taking part in active life in society (Whiteneck et al. 2004). It consists of 25 questions in five dimensions and is well suited to everyday use because of its relatively short form. The instrument mainly captures the physical environment but also asks questions about medical treatment and other services and thus doesn't ignore the social environment. The instrument is primarily designed for subjective self-assessment. Tests have shown that the results are distorted when it is filled out by people who are not directly affected.

Therefore, it shouldn't be filled out by third parties. In order to test the instrument, test persons with various physical limitations were chosen. The reliability ($\alpha = 0.930$) of the instrument was rated high, and so, although it is short, it is suitable for scientific purposes in the area of human-technical interaction.

7. Home and community environment instrument (HACE): This instrument captures the environment at home and in the community (Keysor et al. 2005). It consists of 36 questions in six different dimensions. The questions are answered by grading on a three to five-step scale. The higher the score, the higher the barriers are considered to be that are an obstacle to participation in societal life. The questions cover the physical environment, access to buildings, the condition of the home, as well as aids and so on. But the social environment is also considered, with questions on access to resources such as neighbours and helpful people and other things in the surrounding area. The instrument was developed and tested with people who have difficulties with movement or limitations in communication. In the test phase, the average age of a test person was 70; 60 % were over 65. The researchers concluded that this instrument is well suited to analysing existing barriers for older people in their immediate surroundings. This instrument is also based on subjective self-assessment. For scientific purposes, HACE has not yet been sufficiently tested and so further investigations are needed.

8. Quality of the environment (MQE): This instrument captures individual social and physical environmental factors and barriers (Fougeyrollas et al. 1999). It comprises 109 questions in nine different areas. The questions are answered by ranking in a seven-digit scale: the lower the value, the higher the difficulties gaining access to the element in question. This instrument is also based on subjective self-assessment. There are no statements on quality if the assessment is made by a third party. In testing the instrument, test persons over 60 years of age were used, and the researchers have concluded that this assessment instrument is suitable for analysing the living environment of older people.

9. Collective efficacy: This instrument captures social cohesion among neighbours and targets supportive and obstructive elements in the social environment (Sampson et al. 1997). It comprises 10 questions in two dimensions and so is very short and good for everyday use. Like many instruments already mentioned, it is based on subjective self-assessment. However, there are no statements on whether it is suitable for assessment by a third party. The assessment instrument was tested on a large number of test persons (n = 8,782) in various communities and districts in Chicago, and the reliability ($\alpha = 0.80$–0.91) was assessed as high. Thus, despite its brevity, this instrument is well suited for use for scientific purposes.

10. Community health environment checklist (CHEC): This instrument captures the physical environment of a community (Stark et al. 2007). It was developed to convey the elements in a physical environment that are important for people with movement limitations. On account of this objective, it again derives from the method of subjective self-assessment, as value is placed on the view of the

affected people. The instrument contains 65 questions from different dimensions, which have yes/no answers. Due to this simple answer format, it can be easily integrated into everyday research, despite the high number of questions. It was tested on 19 test persons, most of whom had a stroke. This relatively small number of test persons produced a high validity and reliability result for scientific purposes, but further verification of the quality criteria is indicated.

This overview shows that there is a wide range of assessment instruments for determining environmentally relevant barriers and support factors. It's important to establish early on which target group is to be investigated. Furthermore, it's especially relevant to consider whether the assessment instrument was developed for this very group and validated with representative test persons. Some instruments are specifically designed to cover people with physical functional limitations, for example UIMH, FABS and CHIEF. Other instruments are more focussed on the social environment and cohesion among neighbours. Physical limitations play a less prominent role here, and so the target group is different.

A further point to consider in order to obtain relevant results is the implementation method of the assessment instruments. All the instruments mentioned above were developed for self-assessment. In the case of CHIEF, it is explicitly mentioned that the results may be distorted if it is filled out by another person (proxy) (Whiteneck et al. 2004).

Many assessment instruments were developed in the Anglo-Saxon world, including many of those presented in this chapter. Only the urban identity scale has been published in German. This means that when used in practice the instruments often have to be translated into another language if they are being used outside the Anglo-Saxon world. However, a simple semantic translation doesn't guarantee substantive agreement (Mahler and Reuschenbach 2011).

Translation into another language always means transfer to another culture too. Due to different ideas about the subject in different countries, cultural fine-tuning is necessary. Therefore, it's important to revalidate a translated instrument in the new language. CHIEF, for example, was translated into Chinese, and a new validation was subsequently undertaken (Whiteneck et al. 2004).

In this way, environmental factors that play an important role in the treatment of older people can be analysed and assessed using assessment instruments. At the same time, it becomes clear how important it is to understand the immediate and indirect environment. The persons affected must have the feeling there is improvement in terms of functionality and autonomy through the implementation of AT for them to associate the systems with a clear effect on their situation and an associated (presumed) improvement (de Sousa Leite et al. 2016).

10.4.2 Informal and Formal Carers

Studies have shown that the burden on carers in the home environment is very high. This applies to professional carers as well as informal carers and affects both their physical and mental constitution (de Sousa Leite et al. 2016). AT could potentially help to reduce this burden. Nonetheless, helpful technologies are rarely implemented into practice. There are many reasons for this, ranging from low distribution in healthcare practice to individual rejection by carers. Professional carers in particular seem to be very sceptical about supportive technologies (de Sousa Leite et al. 2016). Among other things, this is due to technical systems not always being implemented in a meaningful and targeted way. The implementation and handling of AT depend fundamentally on the setting and vary according to work flow and task. This complexity clarifies the difficulty in choosing suitable evaluation methods. In addition, the evaluation of AT is often made more difficult by a lack of suitable assessment instruments (Fuhrer et al. 2003). Researchers have therefore tried to develop a conceptual framework for evaluation that doesn't just consider the effectiveness of the technology but also the surroundings and the user. They advocate using assessment instruments—in so far as they are available—and developing new instruments that cover all aspects (Fuhrer et al. 2003).

As early as 2001, Aas argue in a study that there are two different ways to integrate AT into an organisation: the technology is integrated into the organisation or setting, or the organisation adapts to the technology. The degree of acceptance among carers varies according to the option chosen. Furthermore, healthcare organisations—be they for outpatients or inpatients—are very complex and dynamic and therefore require multi-disciplinary analyses. Accordingly, when AT are implemented, consideration should be given to which resources will be used and how the new processes will be organised. Brewster and colleagues (2013) go a step further, investigating which factors are particularly helpful or obstructive from the perspective of carers with regard to the implementation of AT: factors considered particularly positive and supportive were simple and understandable use, opportunities to participate in the development process, flexible usability, opportunities for training and support with problems and simple integration into the work flow. In addition, basic faith in the technology and good support at management level were also indispensable. Factors considered negative or obstructive were possible loss of patient-carer interaction and the patient-carer relationship, and autonomy of carers, as well as a deterioration in the service. Also mentioned were factors such as insufficient training in how to use the system, as well as safety gaps, incorrect data or the technology being unreliable. Interestingly, the possible effects of AT technologies on carers' workload were considered and always rated as particularly obstructive if extra work or no clear reduction in the workload was observed (Brewster et al. 2013). These aspects are particularly interesting given that both formal and informal carers usually work as 'lone warriors' in the domestic sphere.

It is conjectured that the implementation of AT could have a positive influence on people cared for in outpatient settings. However, there are no reliable studies of the situation in Germany in this respect. A broad study in Germany ('Pflege-Thermometer 2016') showed that these technologies had not become established despite their innovative character and the fact that they facilitate greater self-determination in patients. The same applies to the use of telematic and video communication for direct patient contact. These technologies are currently only being used in the areas of work organisation, route planning and internal process improvement (Isfort et al. 2016).

Many assessment instruments exist for measuring burden of care, work organisation and work conditions, or carers' (professional and informal) attitudes to technology. However, few of these instruments are specially tailored to the area of AT. At this point, some selected assessment instruments are therefore referenced that make this connection. In this area, further development of existing assessment instruments and development and testing of new, specific instruments is required. Instruments that can easily be reworked and used in the area of burden of care include the Zarit Burden of Care (Zarit et al. 1985), the Caregiver Strain Index (Robinson and Thurnher 1979) and the Expanded Nursing Stress Scale (French et al. 2000). In the area of organisational factors, there exist the Registered Nurses Working Conditions Barometry index (revised) (Tervo-Heikkinen et al. 2009) and the Revised Nursing Work Index (Aiken and Partician 2000).

One assessment instrument that refers to the quality of care in relation to AT is the KWAZO (Quality of Care; Dijcks et al. 2006). Originally developed for assessment from the client's point of view, this instrument contains seven of the core components of the HEART—Horizontal European Activities of Rehabilitation Technology—criteria (de Witte et al. 1994). However, these components can be transferred to professional and informal carers. Thus, as well as knowing that an AT exists and how to access it, it is essential to know how the technology works in order to make a fully informed decision. All aspects of coordination services are also rated as important in relation to AT. Having a (known) point of contact who coordinates all activities connected with the device (at the micro as well as the macro level) is decisive here. Efficiency and flexibility are further criteria that must be considered. Efficiency refers to lowest possible costs with a high level of control on the side of the service provider, and on the other side a maximum level of service, dependability and as little bureaucracy as possible for the user. Here, there are already indications that the individual attitude (flexibility) to AT plays an intrinsically important role. Devices tailored to the needs of the user promise higher satisfaction levels and greater use. This is equally true of the influence a user has on the technologies developed and used. Higher quality care is associated with a great degree of influence. The final questionnaire contains seven items and makes an assessment on a three-stage scale from insufficient, sufficient to good and is available in English and Dutch (Dijcks et al. 2006).

Demers and colleagues (2009) arrived at similar results. They identified four main categories that should be observed when evaluating AT in relation to carers. Beside stress factors (stressors), these are mainly moderating and mediating factors, but also factors that relate directly to carers, such as quality of life, participation and physical health. For the stressors, the areas used or the form of assistance provided are decisive, but the criteria already mentioned that security, time and level of support should also be borne in mind. As well as role conflicts, prolonged exposure to stressors can lead to very elective time management. At the same time, the domestic environment in which carers have to work (and sometimes also live) contains great potential for stress, especially if it isn't properly designed for using the respective AT. Moderating and mediating factors have a decisive influence on attitudes to the AT—in both directions. For moderating factors, the context plays a decisive role. Duration of care, further support from other informal or professional carers but also the immediate living conditions are all factors that influence the carers' quality of life and health and should be borne in mind. According to Demers et al. (2009) the relationship between the carer and the person cared for is also a factor. At the same time, purely personal 'protection factors' and resources should be considered, because they can also have an influence on individual attitudes.

A further assessment instrument that takes account of the influence of technology in the care context is the CAPSTI-2. A reworking of the first version reduced the instrument to 31 items in four subscales. Those four subcategories are: caring communication, caring involvement, caring advocacy and learning to care. The instrument detects the degree to which the use of technology affects the carer and their caring self-image. However, it has to be adjusted for the outpatient or domestic sector. With the help of this assessment, carers can be divided into three groups (high, medium or low technology influence). In an international sample, the finding was that carers who stated the lowest technology influence also demonstrated the lowest values in the area of direct care activities. Interestingly, the group that indicated a high influence of technology also have the highest values in direct care activities (Arthur et al. 1999). The researchers concluded that where a lot of 'care' is needed little technology is used, and that at the same time where a lot of technology is used a lot of hands on nursing skills are needed. The greater the influence of technology on the relationship between the carer and patient, the more carers have to show their skills, but at the same time act as a kind of mouthpiece when the technology contributes more confusion and uncertainty (Arthur et al. 1999).

10.5 Conclusion

Despite the presumed benefit of AT for people of older age who have physical restrictions or need care as well as for their carers, the use of AT is still very uncommon in Germany. As shown in this chapter, there are a number of reasons for this. Alongside the particular conditions in domestic care settings (particularly in rural regions), which have to be taken into account—but also permanently change —the spread of AT in everyday practice will also depend on current and future research activities. Cooperation among researchers in multi- and trans-disciplinary research associations as well as cooperation among theorists and practitioners should be a goal of utmost importance in this area.

Adding to that is the fact that the people who are themselves affected (or their carers) should be included much more than they have been to date in cooperative development, testing and evaluation. For researchers, this means that more work has to be done regarding information and educational services, but it could lead to higher satisfaction levels among those involved and those affected in the end.

References

Aas IH (2001) A qualitative study of the organizational consequences of telemedicine. J Telemed Telecare 7(1):18–26

Aiken LH, Partician P (2000) Measuring organizational traits of hospitals: the revised Nursing Work Index. Nurs Res 49(3):146–153

Anttila H, Samuelsson K, Salminen A et al. (2012) Quality of evidence of assistive technology interventions for people with disability: an overview of systematic reviews. Technology and Disability 24(1):9–48. doi:10.3233/TAD-2012-0332

Arthur D, Pang S, Wong T et al. (1999) Caring attributes, professional selfconcept and technological influences in a sample of Registered Nurses in eleven countries. Int J Nurs Stud 36(5):387–396

Bartholomeyczik S, Halek M (eds) (2010) Assessmentinstrumente in der Pflege: Möglichkeiten und Grenzen. Schlütersche Verlagsgesellschaft, Hannover

Behrens J, Görres S, Schaeffer D et al. (2012) Agenda Pflegeforschung für Deutschland. Halle (Salle).   Available   via   DIALOG.   http://www.agenda-pflegeforschung.de/AgendaPflegeforschung2012.pdf. Accessed 22 Aug 2016

Beyer H, Holtzblatt K (1999) Contextual design. interactions 6(1):32–42

Bortz J, Schuster C (2010) Statistik für Human-und Sozialwissenschaftler. Springer, Berlin, Heidelberg

Brewster L, Mountain G, Wessels B et al. (2013) Factors affecting frontline staff acceptance of telehealth technologies: a mixed method systematic review. J Adv Nurs 70(1):21–33. doi.10.1111/jan.12196

Brown T, Wyatt J (2015) Design thinking for social innovation. Annual Review of Policy Design 3(1):1–10

Cagney KA, Glass TA, Skarupski KA et al. (2009) Neighborhood-level cohesion and disorder: measurement and validation in two older adult urban populations. J Gerontol B Psychol Sci Soc Sci 64(3):415–424. doi:10.1093/geronb/gbn041

Connell J, Grealy C, Olver K et al. (2008) Comprehensive scoping study on the use of assistive technology by frail older people living in the community, urbis for the Department of Health and Ageing. Available via DIALOG. http://www.health.gov.au/internet/main/publishing.nsf/Content/1D7C7D1598A94CE1CA257BF0001959C2/$File/AssistiveTechnologyReport.pdf. Accessed 12 Feb 2014

de Sousa Leite E, Rodrigues PT, Duarte de Farias MCA et al. (2016) Influence of assistive technology for the maintenance of the functionality of elderly people: an integrative review. International Archives of Medicine 2(21). doi:10.3823/1892

de Witte L, Knops H, Pyfers L et al. (eds) (1994) European service delivery system in rehabilitation technology: a comprehensive description of service delivery systems of 16 European countries. HEART (Horizontal European Activities of Rehabilitation Technology). iRv, Institute for Rehabilitation Research, Line C. Hoensbroek

Demers L, Fuhrer MJ, Jutai J et al. (2009) A conceptual framework of outcomes for caregivers of assistive technology users. Am J Phys Med Rehabil 88:645–655

Dijcks BPJ, Wessels RD, de Vlieger SLM et al. (2006) KWAZO, a new instrument to assess the quality of service delivery in assistive technology provision. Disabil Rehabil 28(15):909–914. doi:10.1080/09638280500301527

Fänge A, Iwarsson S (1999) Physical housing environment: development of a self-assessment instrument. Can J Occup Ther 66(5):250–260

Fougeyrollas P, Noreau L, St. Michel G et al. (1999) Measurement of the Quality of the Environment Version2.0. RIPPH/INDCP, Québec

French SE, Lenton R, Walters V et al. (2000) An empirical evaluation of an expanded nursing stress scale. Nurs Meas 8(2):161–178

Friedrich HF, Eigler H, Mandl H et al. (eds) (1997) Multimediale Lernumgebungen in der betrieblichen Weiterbildung. Luchterhand, Neuwied, Gestaltung, Lernstrategien und Qualitätssicherung

Fuhrer MJ, Jutai JW, Scherer MJ et al. (2003) A framework for the conceptual modelling of assistive technology device outcomes. Disabil Rehabil 25(22):1243–1251. doi:10.1080/09638280310001596207

Gray DB, Hollingsworth HH, Stark S et al. (2008) A subjective measure of environmental facilitators and barriers to participation for people with mobility limitations. Disabil Rehabil 30 (6):434–457. doi:10.1080/09638280701625377

IDEA (1990) Individuals with disabilities education act of 1990. Available via DIALOG. http://uscode.house.gov/statutes/pl/101/476.pdf. Accessed 13 Jun 2016

Intille SS (2013) Closing the evaluation gap in UbiHealth Research. IEEE Pervasive Comput 12 (2):76–79. doi:10.1109/MPRV.2013.28

Isfort M, Rottländer R, Weidner F et al. (2016) Pflege-Thermometer 2016. Eine bundesweite Befragung von Leitungskräften zur Situation der Pflege und Patientenversorgung in der ambulanten Pflege. Herausgegeben von: Deutsches Institut für angewandte Pflegeforschung e. V. (dip), Köln. Available via DIALOG. http://www.dip.de/fileadmin/data/pdf/projekte/Endbericht_Pflege-Thermometer_2016-MI-2.pdf. Accessed 22 Aug 2016

ISO (2010) ISO 9241-210: Ergonomie der Mensch-System-Interaktion: Teil 210: Prozess zur Gestaltung gebrauchstauglicher interaktiver Systeme. DIN Deutsches Institut für Normung, Berlin

Kelley T (2007) The art of innovation: lessons in creativity from IDEO. America's leading design firm. Crown Business, New York et al

Keysor J, Jette A, Haley S (2005) Development of the home and community environment (HACE) instrument. J Rehabil Med 37(1):37–44. doi:10.1080/16501970410014830

Lalli M (1992) Urban-related identity: theory, measurement, and empirical findings. J Environ Psychol 12(4):285–303

Mahler C, Reuschenbach B (2011) Richtlinien zur Übersetzung von Assessmentinstrumenten. In: Reuschenbach B, Mahler C (eds) Pflegebezogene Assessmentinstrumente. Verlag Hans Huber, Bern, Internationales Handbuch für Pflegeforschung und Praxis, pp 101–110

Martin S, Kelly G, Kernohan WG et al. (2008) Smart home technologies for health and social care support. Cochrane Database Syst Rev 8(4):CD006412. doi:10.1002/14651858.CD006412. pub2

Molenbroek J (2013) Putting older people at the heart of every ICT development. In: Mieczakowski A, Clarkson, P (eds) Ageing, adaption and accessibility: Time for the Inclusive Revolution! Engineering Design Centre. University of Cambridge, Cambridge, pp 41–43

Norman DA (2013) The design of everyday things. Basic books, New York

Nutbeam D, Bauman A (2006) Evaluation in a Nutshell. A practical guide to the evaluation of health promotion programs, McGraw-Hill, Maidenhead

Reuschenbach B (2011) Gütekriterien. In: Reuschenbach B, Mahler C (eds) Pflegebezogene Assessmentinstrumente. Verlag Hans Huber, Bern, Internationales Handbuch für Pflegeforschung und Praxis, pp 57–80

Robinson B, Thurnher M (1979) Taking care of aged parents: a family cycle transition. Gerontologist 19(6):586–593

Rosson MB, Carroll JM (2003) Scenario-based Design. In: Jacko JA, Sears A (eds) The human-computer interaction handbook. Lawrence Erlbaum Associates, New Jersey-London, Mahwah, pp 1032–1050

Saelens BE, Sallis JF, Black JB et al. (2003) Neighborhood-based differences in physical activity: an environment scale evaluation. Am J Public Health 93(9):1552–1558

Sampson RJ, Raudenbush SW, Earls F (1997) Neighborhoods and violent crime: a multilevel study of collective efficacy. Science 277(5328):918–924

Stark S, Hollingsworth HH, Morgan KA et al. (2007) Development of a measure of receptivity of the physical environment. Disabil Rehabil 29(2):123–137. doi:10.1080/09638280600731631

Stockmann R, Meyer W (2014) Evaluation. Eine Einführung, 2nd edn. Barbara Budrich UTB, Opladen-Toronto

Tervo-Heikkinen T, Kiviniemi V, Partanen P et al. (2009) Nurse staffing levels and nursing outcomes: a bayesian analysis of finnish-registered nurse survey data. J Nurs Manag 17 (8):986–993. doi:10.1111/j.1365-2834.2009.01020.x

Topo P (2009) Technology studies to meet the needs of people with dementia and their caregivers: a literature review. J Appl Gerontol 28(1):5–37. doi:10.1177/0733464808324019

Whiteneck GG, Harrison-Felix CL, Mellick DC et al. (2004) Quantifying environmental factors: a measure of physical, attitudinal, service, productivity, and policy barriers. Arch Phys Med Rehabil 85(8):1324–1335. doi:10.1016/j.apmr.2003.09.027

Zagler WL (2013) Rehabilitationstechnik—assistive technologie. In: Fialka-Moser V (ed) Kompendium Physikalische Medizin und Rehabilitation. Springer-Verlag, Wien, Diagnostische und therapeutische Konzepte, pp 245–258

Zarit SH, Orr NK, Zarit JM (1985) The hidden victims of alzheimer's disease: families under stress. New York University Press, New York

Chapter 11
Assistive Technology for People with Dementia: Ethical Considerations

Hesook Suzie Kim

Abstract Assistive technology (AT) for people with dementia is in general used to improve the quality of life and prolong independent living in the community as long as possible by (a) compensating for memory impairment and disorientation, (b) helping to ensure safety which is oriented to protecting persons to remain at home safely, and (c) improving emotional status and decreasing behavioral problem which are oriented to manage psychological and behavioral problems. The AT use for people with dementia, especially those AT used for safety such as monitoring/tracking devices, raises various ethical issues related to the values of autonomy, personal dignity, privacy, and personhood. The decisions regarding the use of AT devices and implementing their uses thus have to follow a systematic examinations of various ethical issues from the perspectives of the person with dementia, family caregivers, and professional providers in order to ensure that the autonomy, privacy, and personhood are upheld at the same time achieving the highest level of safety and comfort possible for the person with dementia with the AT use.

Keyword Assistive technology · Dementia · Alzheimer's disease · Ethical dilemma · Monitoring/tracking devices

11.1 Introduction

The focus of this chapter is on addressing key ethical issues associated with the use of assistive technology (AT) for people with dementia. Dementia is a neurocognitive condition that mostly affects the elderly and is characterised by a decline in cognitive function influencing the ability to perform everyday self-care activities. It is represented by a decline, usually progressive, in memory, problem-solving,

H.S. Kim (✉)
Institutt for Helsefag, University College of Southeast Norway, Drammen, Norway,
Professor Emerita, College of Nursing, University of Rhode Island, Kingston, RI, USA
e-mail: hsuziekim@comcast.net

language and other cognitive skills that affect a person's ability to perform everyday activities and function independently and competently in daily life.

It is estimated that Alzheimer's disease (AD) accounts for about 60–80 % of dementia cases, and one in nine people aged 65 over in the USA currently has the disease (Alzheimer's Association 2016a). Dementia and AD are progressive, moving from the early stage of relative self-sufficiency to the late stage in which people become completely dependent. The Alzheimer's Association categorises the disease's progression in three stages—the early stage of mild AD, the middle stage of moderate AD and the late stage of severe AD—and gives the following examples of common difficulties typical of each stage (Alzheimer's Association 2016b):

Early stage (mild AD)

- Problems coming up with the right word or name
- Trouble remembering names when introduced to new people
- Having greater difficulty performing tasks in social or work settings
- Forgetting material you have just read
- Losing or misplacing a valuable object
- Having increased trouble with planning and organising.

Middle stage (moderate AD)

- Forgetting events or your own personal history
- Feeling moody or withdrawn, especially in socially or mentally challenging situations
- Being unable to recall your own address or telephone number or the high school or college from which you graduated
- Confusion about where you are or what day it is
- Needing help choosing proper clothing for the season or the occasion
- Trouble controlling bladder and bowels in some individuals
- Changes in sleep patterns, such as sleeping during the day and becoming restless at night
- Increased risk of wandering and becoming lost
- Personality and behavioural changes, including suspiciousness and delusions or compulsive, repetitive behaviours such as hand-wringing or tissue shredding.

Late stage (severe AD)

- Requiring full-time, round-the-clock assistance with daily personal care
- Losing awareness of recent experiences as well as of your surroundings
- Requiring high levels of assistance with daily activities and personal care
- Experiencing changes in physical abilities, including the ability to walk, sit and, eventually, swallow
- Having increasing difficulty communicating
- Becoming vulnerable to infections, especially pneumonia.

Because dementia is progressive, there has been a push not only by people with dementia and their carers but also by the healthcare sector to keep people with dementia in the community, delaying their placement in long-term care institutions for as long as possible. Keeping people with dementia in the community and in their own home setting, usually during the early and middle stages, requires support to keep people as independent as possible and safe. Sustaining independence and ensuring safety have become the hallmarks of good community care for people with dementia, and in the last 10 years AT has become an important resource for attaining those hallmarks.

In their report, O'Keefe et al. (2010) list three categories of AT that are useful for people with dementia at home:

1. devices that compensate for memory impairment and disorientation
2. devices that help to ensure safety, which are designed to protect people so they can remain at home safely
3. devices that improve emotional status and decrease behavioural problems, which are designed to manage psychological and behavioural problems.

AT that compensates for memory impairment and disorientation can be classified into two groups. The first includes mostly electronic devices for finding frequently lost or misplaced items, making and receiving phone calls, providing medication reminders, reminding and prompting, and orienting people as to the month, date and time. The report states that as the use of such 'reminder' electronic devices requires people to have intact executive cognitive function, which is often deficit in persons with dementia, their use may be limited. However, it is suggested that the use of such devices by persons with dementia in the early stage may establish behavioural routines that can be sustained even with the progression of cognitive deficit.

The second group of AT is designed to help ensure safety for persons with dementia either by alerting carers to unsafe situations or preventing unsafe incidents and situations. Devices in this group are more actively oriented towards addressing problems that arise from increasing cognitive impairment in relation to memory and decision-making. This group of devices includes devices for car safety, summoning help, preventing injury at home, monitoring activity in the home and helping to ensure safer walking and prevent wandering. Although there are ethical implications associated with the use of AT for persons with dementia in general, the use of this second group of AT raises critical ethical issues as their use impinges on privacy, personal integrity, autonomy and personal freedom.

An extensive survey by Gibson et al. (2016) of AT available in the UK for people with dementia identified three types:

1. AT used 'by' people with dementia such as time/place orientation devices, prompting and reminder devices, communication aids, dementia-friendly tools for assisting with the activities of daily life and alerts/alarms
2. AT used 'with' people with dementia including communication aids, game and play devices for entertainment, and devices for reminiscence

3. AT used 'on' people with dementia including telecare systems, Global Positioning Satellite (GPS) and location tracking devices, safety and security devices such as geofencing products, telephone blockers and key safes.

Alzheimer Europe (2012) states that AT for people with dementia should be 'assistive' to the person with dementia and lists the potential benefits of AT for people with AD as:

1. enabling/promoting autonomy and well-being
2. providing protection and a safe environment
3. contributing towards quality of life and social inclusion
4. providing benefits to professional and/or residential care.

The development of new AT for people with dementia has been active in recent years. Sophisticated sensor systems, GPS system applications, information technology for interactive interface systems, intelligent technology and robotics aim to keep people with dementia in the community as long as possible and to improve their quality of life even with advanced cognitive impairment. New developments certainly benefit people with dementia, and personal and professional carers, by improving safety and quality of life, and delaying institutionalisation. But at the same time, there are additional ethical implications in using such advancing technologies.

11.2 The Types and Characteristics of AT for Safety

AT for safety for people with dementia is basically designed to protect people from harm by:

1. directing the people to perform safe practices through the use of 'reminder' technology and communication aids
2. preventing people from engaging in potentially injurious actions with the use of alarm/alert devices and communication devices
3. monitoring people's activities and movements in order to foresee or prevent harm to them with the use of tracking/surveillance devices.

Various 'reminder' technologies are useful to ensure people perform necessary tasks, such as taking medication, and follow through activities, such as personal hygiene and preparing meals, and to prevent people from engaging in unsafe activities, such as leaving home at night or leaving the gas cooker on. However, as people with dementia become increasingly forgetful, are at increasing risk of falling and are unpredictable in their movements, and wander and get lost, such 'reminder' technologies alone often become inadequate to protect them from harm. Thus, the main features of AT that specifically ensure safety and prevent injury and harm (both physical and emotional) are monitoring and surveillance. Home monitoring, movement monitoring and surveillance devices collect information about people's activities, movements and locations, providing real-time information to carers and professional care-providers.

There are two types of electronic monitoring and surveillance devices: active and passive. Active devices require actions or responses from the person being monitored, while passive devices do not. An active device requires people to recognise that they need help in a situation and summon the necessary help by actions such as pushing a button on a device, dialling a pre-programmed number on a mobile phone or speaking into a communication system set-up for the device. However, using such devices requires a degree of executive cognitive function, and in general people with dementia beyond the early stage would not be able to accomplish the necessary tasks.

Passive technologies do not require active involvement by the person being monitored, and come in two types: environmentally oriented and person-oriented. Examples of environmental monitoring technologies are heat and gas detectors, water temperature monitors, and room air temperature monitors, which monitor and control environmental conditions to ensure they remain within a safe range. By contrast, person-oriented passive monitoring devices monitor people, checking, for example, their movements, activities and locations. There are many wearable devices that transmit information about people's movements, activities and whereabouts both while they are in their own homes and away from their homes. Such devices are especially helpful in ensuring people remain safe when they are not with carers. Since a large proportion of people in the early stage of dementia remain in their homes alone or are with carers for only a portion of the day, passive monitoring has become essential for ensuring people remain safe while alone.

One of the most rapidly developing technologies for passive monitoring is location-tracking devices interfaced with the web using the GPS systems. In recent years, such technologies have allowed important advances in dealing specifically with the problem of wandering and 'getting lost' in people with dementia when they are not in their homes. A variety of such devices is available on the market in the US (Alzheimers.net 2014). Wandering and getting lost are prevalent in non-institutionalised people with dementia. For example, in a study of 218 people with dementia living in the community, 33–40 % report getting lost during a period of 2.5 years (Pai and Lee 2016), and wandering and getting lost are one of the main reasons why people with dementia are placed in long-term care institutions.

The purpose of tracking/surveillance devices is to keep people with dementia safe by using information about their location to make it possible to intervene before they become lost. AT for increasing safety in persons with dementia, thereby also alleviating anxiety in carers, is considered a critical technology for dementia (Rialle et al. 2008). However, controversies remain regarding the use of such GPS-enabled tracking systems, and there are two opposing attitudes to their use. Advocates claim they ensure people remain safe and prevent people with dementia from coming to harm while allowing them freedom of movement and thereby maintaining their autonomy. Critics believe their use stigmatises and dehumanises people, invading their personal privacy and threatening their personal autonomy and integrity.

11.3   Ethical Considerations

11.3.1   Scenario

Anne Williams is a 72 year-old woman living alone in a large city on the west coast of the US in a housing complex for elderly people. She has been a widow for 10 years, and her daughter is her only close relative. Her daughter lives about 65 miles away and was given full power of attorney and durable power of attorney for Anne's healthcare 6 months ago after Anne got lost in the city for several hours and was escorted home by the police. Following the incident, she was diagnosed with Alzheimer's disease at the late-early stage. Although her daughter had suspected that Anne had Alzheimer's and had asked her to be diagnosed during the previous year, Anne had resisted revealing her problems to her doctor until this incident happened.

Anne has the typical signs and symptoms of the late-early stage of AD. She has difficulty finding the right words and in general food preparation, and she forgets where she has put valuable and important items, such as legal documents and cash. She gets lost and has difficulty recalling recent activities or experiences. She talks about experiences from the past as if they had happened recently and she is suspicious of people. She was referred to a day care centre for elderly people for five days a week, which she has been attending.

Her daughter is not able to care for Anne as she has three young children aged 4, 10, and 12, and works part-time. Her husband travels to out-of-state locations a great deal because of his work, and one of the children attends a special programme for children with learning difficulties. Anne asked her daughter to be her formal guardian to help her to manage personal, financial and health matters, but has insisted that she will stay on her own, living in her own apartment. Since she is not at the stage of illness that requires placement in a total-care institution, she continues to live on her own in the apartment. Her daughter asked her if she would agree to have a home health aide come to her apartment when she is home alone, but Anne adamantly refused, claiming she could manage on her own. She insists she will be fine by herself, although she has stopped preparing her own meals and seems to have forgotten how to cook. She relies mostly on pre-packaged foods. She gets around on public transport and on foot. She has got lost a couple of more times since the incident, getting off the bus at the wrong stop, but was able to find her way home eventually. She has a very small social circle and is on her own most days, except when she is attending the day-care centre. Her daughter is only able to visit her a couple of times a month, usually spending few hours with her checking that everything is okay, doing the laundry, delivering food and supplies and setting up the reminder board.

As Anne usually has breakfast and lunch at the day-care centre, her dinners are light, consisting usually of yoghurt, fruit cup and a sandwich. On weekends, her daughter calls at breakfast and lunch to give her step-by-step instructions over the phone on getting the meals ready mostly by using the microwave and an automatic shut-off kettle. The gas supply for the range and oven has been turned off after several incidents of pots and pans getting burnt. Anne's daughter has also installed several devices so that Anne can be safe at home including: a dementia clock, a medication reminder for weekends when she is not at the day-care centre set up with the help of day-care centre staff, a front door sensor that is wired to a board on the door with lighted messages that remind Anne to take her keys, mobile phone and ID with her when she leaves the apartment, and a lowered water temperature for the hot water supply system. Anne also wears a MedicAlert bracelet with an enrolment in the Alzheimer's Association SafeReturn.

Upon hearing about the additional times Anne had been lost, her daughter became agitated about her getting lost again and thought that the MedicAlert bracelet might not be enough to protect her. This prompted Anne's daughter to investigate GPS-enabled tracking units to monitor Anne's location. When her daughter asked Anne if she would be willing to carry the device so that she would know where she was, Anne agreed that she would. With her consent, a device was purchased from a company called Pocket Finder and a monthly monitoring service was contracted for Anne. Anne's daughter explained how the device works, that it should be charged at least every other day, and that Anne should carry it along with her mobile phone whenever she leaves her apartment. (Before investigating GPS-enabled tracking systems, Anne's daughter looked into the possibility of installing location services on Anne's mobile phone, but found that the model Anne has is not configured for this. She then considered getting a smart phone instead of a tracking system, but decided against it as she knew how Anne used her mobile phone. Anne knows how to make phone calls on her current mobile phone to three individuals only using the programmed speed dial and returns calls following the mechanically remembered steps. Anne's daughter was convinced that it would be impossible for Anne to learn to use a smart phone.) Anne stopped carrying the tracking device which is the size of a small pager after five days, and it sits on top of her bureau by the charger. When asked why she wasn't carrying it, she insisted that she could find her way home okay and didn't need it. However, her daughter is worried that one day Anne will get lost and she won't know where she is, though so far she has not insisted that her mother carry the device.

11.3.2 Dilemma

The dilemma in this situation is how to keep Anne in the community living in her home as safely and independently as possible without impinging on her autonomy and integrity as a person. The question is: what are the rights of the person with dementia and the obligations of carers in such situations with regard to using the AT of monitoring/tracking systems? There are other questions too. What are the reasons for Anne's decision not to use the device? Should Anne's decision not to use the device be respected? What are the short-term and long-term implications of using and not using the device on Anne and her daughter? What next steps could Anne's daughter take to protect her mother?

AT for people with dementia, especially monitoring/tracking devices, may ensure people with dementia remain safe, enabling them to live more freely and independently and thus promoting their autonomy, but it may also be seen as intruding on privacy, causing stigma and stigmatisation and as jeopardising individuals' autonomy by controlling them. As monitoring/tracking technology stores and shares information about the person being monitored (such as where they are and what they are doing), the use of this AT impinges on the monitored person's privacy. As monitoring/tracking or tagging systems are used in the criminal justice system in many countries, in recent years especially for tracking sex-offenders and others who are considered to be dangerous, there is a stigma attached to such systems, which can make the people using them feel stigmatised and can lead to stigmatisation by others. Although the tracking device used in this scenario requires the person with dementia to participate in its use by carrying the device, there are devices which are operated without active engagement by the person with dementia, such as shoes or clothing with embedded tags or sensors, which can be used without the consent or knowledge of users.

Nuffield Council on Bioethics (2009) offers an ethical framework for caring for people with dementia, which can be used as a basis for examining ethical issues associated with using AT for people with dementia who live in the community. The ethical framework, which aims to establish an approach to care based on respect for the needs, preferences and personhood of the individual with dementia, has six components:

Component 1: A 'case-based' approach to ethical decisions: Ethical decisions can be approached in a three-stage process: identifying the relevant facts; interpreting and applying appropriate ethical values to those facts; and comparing the situation with other similar situations to find ethically relevant similarities or differences.

Component 2: A belief about the nature of dementia: Dementia arises as a result of a brain disorder, and is harmful to the individual.

Component 3: A belief about quality of life with dementia: With good care and support, people with dementia can expect to have a good quality of life throughout the course of their illness.

Component 4: The importance of promoting the interests both of the person with dementia and of those who care for them: People with dementia have interests, both in their autonomy and their well-being. Promoting autonomy involves enabling and fostering relationships that are important to the person, and supporting them in maintaining their sense of self and expressing their values. Autonomy is not simply to be equated with the ability to make rational decisions. A person's well-being includes both their moment-to-moment experiences of contentment or pleasure, and more objective factors such as their level of cognitive functioning. The separate interests of carers must be recognised and promoted.

Component 5: The requirement to act in accordance with solidarity: The need to recognise the citizenship of people with dementia, and to acknowledge our mutual interdependence and responsibility to support people with dementia, both within families and in society as a whole.

Component 6: Recognising personhood, identity and value: The person with dementia remains the same, equally valued, person throughout the course of their illness, regardless of the extent of the changes in their cognitive and other functions.

(Nuffield Council on Bioethics 2009, p. 21).

The Alzheimer's Society of Canada (2011) also specifies that a person-centred philosophy, which encompasses dignity and respect, information-sharing, participation and collaboration as its core values, should be the basis for the care of people with dementia. The commonly upheld ethical values for healthcare including autonomy, beneficence, non-maleficence and justice can also be applied to the care of people with dementia in general and to the use of AT in particular (Alzheimer Europe 2012). The Alzheimer's Society of the UK (2014) recommends ethical guidelines that consist of fair access to technological solutions, avoiding unintentional harm, respecting privacy and confidentiality, and ensuring data security in the use of AT for people with dementia. In their position paper, the Alzheimer's Society of the UK (2011) also raises concerns about the use of AT for people with dementia in terms of the risk of social exclusion, the potential threat to independence, the danger that people's lives and living environment will be made more complicated, the danger that the focus is on a person's problems not their strengths, which can damage self-esteem, the danger of AT being applied without the full consent of the person with dementia, data protection issues and stigma.

There are two sets of ethical issues in the use of monitoring/tracking devices (and AT in general) for people with dementia that are interrelated: first, issues associated with the use of the devices, and second, issues regarding who should make decisions about the use of the devices. The main ethical issues in the use of monitoring/tracking devices in people with dementia are their potential to violate individuals' civil rights, impinge on their privacy, freedom and autonomy, limit their self-determination in taking risks, and create feelings of helplessness (O'Keefe et al. 2010; AtDementia 2007), which are weighted against the benefits of their use. The ethical board for the MINAMI project established a set of ethical guidelines for developers and designers of mobile-centric ambient intelligence in health and homecare, which included privacy, autonomy, integrity and dignity, reliability, e-inclusion and benefit to society as the key principles (Kosta et al. 2010) that are appropriate in the use of AT for dementia care. These guidelines specify that privacy means an individual should be able to control access to their own personal information and to protect their own space, autonomy means they have the right to decide how and for what purposes they use technology, and the principle of integrity and dignity requires individuals to be respected and their dignity as human beings to be maintained when technical solutions are applied. Similarly, the ethical guidelines applied in the EU-funded research and development technology project, Assisting family Carers through the use of Telematics Interventions to meet Older persons' Needs (ACTION), are based on the ethical principles of respect for human dignity, worth and fundamental rights, autonomy and privacy, confidentiality, informed consent, non-maleficence, justice, beneficence, and veracity or truth-telling (Magnusson and Hanson 2003).

Maintaining integrity and personhood when adopting AT for people with Alzheimer's disease is in line with the critical ethical principles of respect encompassing the concepts of autonomy, self-determination, personal dignity, beneficence focusing on protecting the person, non-maleficence in preventing harm, and justice with regard to equal treatment regardless of vulnerability. The principles highlight the need to address both the value of AT devices in assuring better outcomes (that is, quality of life and experience) for people with dementia in terms of preventing harm/risks and their affects in producing other physical, emotional, and social responses in the people, as well as the meaning the devices have for people with dementia and their carers. The benefits of monitoring/tracking devices for people with dementia are that they provide them with a sense of freedom to walk where they wish (O'Keefe et al. 2010) and protect them from potential harm while not restricting their routines (Alzheimer's Society 2014). The use of such devices does not have physical consequences as the devices are usually small and carried on the person as bracelets/anklets/pendants or as a part of clothing or shoes. The emotional and social effects of the devices are a controversial topic. On the one hand, using such devices may provide people with a sense of safety, feelings of autonomy and freedom, and opportunities to engage in normal social contacts and activities. On the other hand, using these devices may make people feel stigmatised or that their privacy has been invaded. In a systematic review of the literature from 1990 to 2009 regarding the ethics of using AT for people with dementia, Zwijsen et al. (2011) found three major contexts for ethical concern. The themes of privacy, autonomy and obtrusiveness were identified in the context of the personal living environment, that is private homes, while the themes of stigma and human conduct were found in the context of the outside world, as AT is used in interaction with the outside world. The themes of individual approach, affordability and safety were identified in the context of the design and application of the device. The literature indicated various sorts of controversies regarding these themes:

1. Privacy—concerns regarding privacy in relation to safety; privacy in relation to data availability; tendency for carers to be more concerned about privacy than the people with Alzheimer's.
2. Autonomy—self-determination in maintaining autonomy as the major issue; possibility of control and dehumanisation; possibility of coercion to use AT.

3. Obtrusiveness—inexact definition of the term's use but mostly referring to AT being undesirably prominent or noticeable; context specificity; degree of intrusiveness and consequent adoption usually influenced by prior experience, others' attitudes, the characteristics of the device and their effects.
4. Stigma—AT use indicative of dependence, meaning users and their families may feel stigma and be stigmatised.
5. Human conduct—potential for human contact to be reduced by the use of AT or conversely for human contacts to be stimulated.
6. Individual approach—desirability of an individual approach in the design and application of devices.
7. Affordability—possibility that the principle of justice is violated by the expense.
8. Safety—influence of the characteristics of the devices themselves or the ways they are used on ensuring safety.

In general, people with dementia seem to have a positive attitude toward using AT if it facilitates their independence and autonomy (Robinson et al. 2009). However, the use of AT by people with dementia often depends on decisions made by those people together with their carers and significant others. In their interview study of nine pairs of people with dementia and their family carers, Smebye et al. (2016) found major ethical dilemmas in the use of AT such as:

1. the conflict between the autonomy of the person with dementia and the need to prevent harm, which was viewed as important by the carer (i.e. non-maleficence)
2. the conflict between the autonomy of the person with dementia and the approach advocated by the family carer as promoting well-being (i.e. beneficence)
3. the conflict between the autonomy of the person with dementia and the autonomy of the family carer.

In an intervention study of 10 participants about incorporating the use of AT in the daily lives of persons with dementia, Lindqvist et al. (2013) found that the use of AT was mostly utilitarian. The decision to use AT was taken because of the need to maintain valued activities, and use was continued because the capacity to use AT and the feeling of control developed with its use, usually with the support of significant others because of AT's value to them. In their follow-up study of the participants in the earlier report (Lindqvist et al. 2013), Lindqvist et al. (2015) found that several features of the AT used, such as availability of constant visible information, personalisation, feedback features, required user involvement, and appropriateness to the physical context and activity, influenced whether user goals, such as efficiency, effectiveness and safety, as well as a sense of control, were achieved. On the other hand, in a study of cognitively intact older adults asking their views on AT use, Landau et al. (2010) found the key reasons are that it is used to be for carers' peace of mind, for the sake of the safety of the person with dementia, for specific needs and for maintaining the person with dementia's autonomy.

The respondents raised issues regarding conflicts between respecting the consent and decisions of the person with dementia versus keeping the use secret, and between protecting privacy versus the possibility of decreasing risks, which were seen as justification for AT use. Some respondents thought that when people with dementia lack cognitive competence their rights to privacy and freedom of movement may be denied. There are indications that people with dementia often used AT not because they believed that the devices benefited them but because their carers wanted them to use them (Gibson et al. 2015). Persons with dementia seem to be better disposed to using AT if they are able to fit the devices into their lives and find them useful. On the other hand, carers considered 'safety' to be the primary aim, with safety concerns overriding concerns about privacy or restriction of the person with dementia's integrity (Rosenberg et al. 2012). Godwin (2012) also indicates that carers considered 'autonomy-promoting' devices to be ethical and acceptable, while Gibson et al. (2015) found that carers considered using AT often to be beneficial in providing support with their care-giving tasks, beneficial in giving them 'peace of mind' and offering greater independence to the person with dementia. In addition to conflicts stemming from differences in the reasons for using AT, conflicts are possible in relation to who makes the decision to use AT since the main reasons for using it are perceived differently by people with dementia and carers. Landau et al. (2011) found in a survey of respondents composed of cognitively intact elderly people, family carers of people with dementia, social workers and other professionals, and social work students that the group as a whole considered the most appropriate person to make the decision to use a GPS-based tracking device to be the spouse of the person with dementia, the family member most involved and the person with dementia in that order. Such findings suggest possible conflicts when there is a difference between the desires of people with dementia and those of their spouses or family carers.

Nuffield Council on Bioethics (2009) offers the following ethical guidelines for using AT in people with dementia:

- A person with dementia's decision to accept or refuse use should be respected if the person has the capacity to make choices.
- In cases where the person with dementia lacks the capacity to make an independent choice, decisions should be on a case-by-case basis, relying on (a) the person's past and present views and concerns, (b) the benefits of the AT, (c) the extent to which a carer's interests may be affected and (d) the dangers of loss of human contact.
- There should be an effort to balance freedom and risk through a risk-benefit assessment incorporating considerations about the well-being and autonomy of the person with dementia, as well as their need for protection from physical harm and the needs and interests of others.

A report by Alzheimer Europe (2012) presents the main ethical issues in the use of tracking/tagging devices as conflicts between autonomy and liberty versus safety/security, allocation of responsibilities, privacy, dignity and personhood, and stigmatisation. These concerns are summarised here:

1. Conflict between autonomy and liberty, and safety/security: conflicts may arise in decision-making because of differences in attitudes between people with dementia and carers in balancing respect for a person's right to autonomy and liberty with the perceived need for safety.
2. Allocation of responsibilities: allocation of responsibilities regarding who is responsible for making decisions, for the use of AT itself and for safe-guarding information obtained by using AT is not well defined so that people's privacy is protected and that AT is equitably available to people who need it.
3. Privacy: potential invasion of privacy has been raised as an issue in the use of monitoring/tracking devices as the person using the device may not be able to escape from being monitored.
4. Dignity and personhood: the use of monitoring/tracking devices is viewed as a violation of civil liberties by some, who object to such devices being used to monitor people with dementia. They view the use of such devices as dehumanising for the person, restricting their personal liberty and stripping away their personal dignity. Conflict in this regard may exist not only among family carers, who participate in decision-making around safety issues, but also among people with dementia, who have to weigh up the benefits of freedom of movement against the loss of dignity associated with using monitoring devices.
5. Stigmatisation: the stigma attached to using tracking/monitoring devices adds to existing stigmatisation around dementia. Using these devices may be stigmatising for the user, prompting feelings of incompetence and helplessness, even if the device is unobtrusive. Others may also stigmatise the user, labelling them as dependent and incompetent, thus dehumanising and objectifying them. Since stigma and stigmatisation are social in nature, this may also influence the ways in which device users engage in social interactions.

Ethical issues and considerations addressed in this section on the use of AT by persons with dementia with a specific focus on the use of monitoring/tracking devices highlight the complexity associated with making the decision to use these devices and in their actual use. Often ethical concerns and conflicts are individual and context-oriented. The ways such conflicts influence people with dementia and their carers vary, ranging from non-adoption, quitting use after a trial, continued use with dissatisfaction and unhappiness, and satisfied and content use. Such results are often associated with the attitudes of the people involved (usually the person with dementia and the carers) and the processes by which both the decision-making and implementation occur. Device use can also be efficacious or inefficient as well as effective or ineffective in influencing outcomes, and this is often due to the characteristics of the AT devices themselves or the manner in which they are used. A framework and process are needed to ensure successful and satisfactory use of AT by people with dementia.

11.4   Recommendations and Conclusions

People with dementia living in the community differ greatly in terms of their clinical, cognitive and behavioural capacities and the resources available to them to remain successfully in the community. More than one third of this population is believed to be living alone, and many require care support and resources from significant others and community services. As dementia is progressive and most of the medical management for people with dementia is not therapeutic but supportive, the emphasis has been on ways to sustain as much independence as possible for as long as possible, with safety being the utmost concern in the context of cognitive impairment. The use of AT has become the fastest-growing way to reach this objective. Although many view the most critical consideration for using AT in people with dementia as the question of who should ultimately benefit from the use, there are many ethical issues surrounding the use itself and the decision to use AT, as discussed in the preceding section. In addressing these ethical concerns, many guidelines have been proposed to ensure appropriate, efficacious and effective use of AT (Bjørneby et al. 1999; Marshall 2000; O'Keefe et al. 2010; Alzheimer Europe 2012; Alzheimer's Society 2014). Based on the findings from her qualitative study regarding ethical issues surrounding the use of AT, Godwin (2012) also developed six questions as guidelines for making ethical decisions on using AT applying an individualised approach:

1. For whose benefit is the equipment being used?
2. Whose definition of benefit is being applied?
3. What are the potential affects of the technology on the well-being of the individual and their carer? Does it support the person's autonomy or simply reduce risk?
4. What are the actual and potential active and passive detrimental affects of the technology?
5. What are the costs and benefits (physical, emotional, psychological, ethical and financial) of using the technology—and to whom do they apply?
6. What is the real (not hypothetical) alternative to using this piece of AT?

A person-centred approach has been advocated as a framework for the proposed guidelines (Woolham 2006; Bonner and Idris 2012; Alzheimer's Society 2011). A person-centred approach in healthcare is based on a philosophy that upholds a person's individuality and uniqueness and their right to self-determination and personal respect, and favours individual tailoring of care. The guidelines proposed by Bjørneby et al. (1999), published in *Technology, Ethics and Dementia (TED): A guidebook on how to apply technology in dementia care,* are a tool for practitioners in dementia care to make ethical decisions on the use of technology. The guidelines have been applied in an intervention study to encourage the use of AT by people with dementia (Lindqvist et al. 2013) and are recommended for professional practitioners. However, they can be adopted as a framework that incorporates a person-centred philosophy for individualising the use of AT by people with

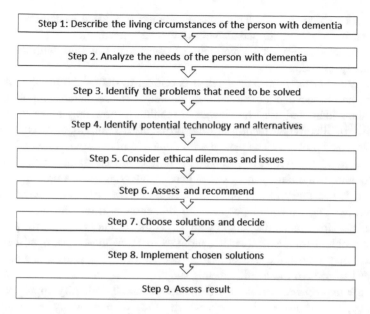

Fig. 11.1 Steps in the use of technology for persons with dementia, adapted from Bjørneby et al. (1999)

dementia living in the community in relation to decision-making and the use of AT involving the people with dementia themselves, personal and professional carers, and/or significant others. The guide includes the following nine steps (Fig. 11.1).

The steps in the illustration incorporate the guidelines specified in TED (Bjørneby et al. 1999) and recommendations for individualising decisions on using AT gleaned from the literature.

1. *Describe the living situation of the person with dementia*: Living situation doesn't just refer to whether a person lives alone or with a spouse or others, but also to factors such as what kinds of help and support the person uses for their needs and what kinds of daily or regular activities the person engages in, such as shopping for food, attending a day-care centre, participating in regular social activities, such as a recreation club, and attending a church. A person's daily habits, particularly with regard to what they do outside the home and how much they use everyday technology, such as the TV, a mobile phone and computer, provide critical information about the person's usual pattern of life.

2. *Analyse the individual needs of the person with dementia*: TED emphasises that the focus should be on analysing individual needs rather than problems. Analysing individual needs means considering them from the perspective of the person with dementia. The priority has to be what the person themselves considers to be their most important needs in relation to daily life. Going for a daily walk may be an important need and a desire for one person, while a long-standing habit of going out for a meal at a restaurant may be an important

need for another person. This analysis should provide a systematic and comprehensive view of the person with dementia's needs from their specific perspective, as an individual's hierarchy of needs may not align with the standard hierarchy of needs.

3. *Identify the problem to be solved*: It is critical to define the problem and situation, and associated functional requirements, as completely as possible. The problems identified must be viewed in relation to how they affect the person's quality of life. When identifying the problem, it's important to consider who views it as a problem. The person with dementia may not view something (such as getting off at a wrong bus stop and getting confused about where they are) as a problem, while family members or carers may consider it a problem, in which case it is necessary to find out why such a disparity exists and what it means. Cognitive impairment in people with dementia is often the reason for such conflicts. However, if the person with dementia does not consider the problem to be a problem for them, then it continues to be a non-problem in their mind, which may render the use of AT unsuccessful or ineffectual. Therefore, it is critical to resolve the conflict at this stage.

4. *Identify appropriate technology and alternatives*: To identify appropriate technology, it is necessary to assess which kinds of technology are available to address the problem identified in the person with dementia's living situation. Identified technologies should be assessed in terms of how well they meet the needs of the individual in terms of upgrading their quality of life and resolving the problem, and at the same time how they affect other aspects of the person's life. The value of the technology has to be considered in terms of beneficence and non-maleficence as well as in terms of unintended maleficence. It should also be assessed in terms of the required involvement of the person using it (i.e. responses/activities such as pushing a button, making a phone call, choosing an action, carrying the device and charging a device regularly). The complexity of such involvement has to be considered in light of the person's capacities and willingness to use the technology. Assessment of different kinds of technology should also address the ethical implications of devices in terms of how they may impact on the person with dementia's privacy, integrity and autonomy.

5. *Discuss ethical dilemmas and issues around potential technology and alternative solutions*: When making decisions it is important for carers and family members to consider possible ethical dilemmas and issues, and this requires these surrogate decision-makers to engage in reflective dialogue with oneself (understand one's own perspectives, attitudes etc.) and discuss the issues with the person with dementia, professionals and others in order to free themselves from their own wishes, biases and positions about the 'true' features of the AT and the motivation for using it. When they've done this, it is possible to resolve conflicts from the perspective of the person with dementia and incorporate technology that is considered 'absolutely necessary' into the situation.

6. *Assess and recommend technology*: Each type of technological device identified in the earlier steps should be evaluated according to five criteria to determine its applicability. The criteria include: (a) a clear, documented need for a technical

solution, (b) a positive improvement in the person with dementia's safety, security, health or independence in using the technology, (c) satisfaction of user needs through this improvement, (d) least possible infringement on user integrity, and (e) clear documentation of procedures. Potential technology should meet all of these criteria in order to be considered appropriate for use in a specific situation. Another issue at this stage is finance. If the cost of the technology has to be covered from the person with dementia's personal resources or those of their family, an expensive device may be considered less preferable even if it meets the criteria in a more comprehensive manner. If the technology is paid for by public funds through health insurance or government programmes, choices may be limited. Communities make various public and private sources of financial assistance available for technology, and decision-makers (persons with dementia, family members and carers) need to be aware of these.

7. *Chose solution and decide*: Ethical issues need to be resolved before a specific decision is made to use an AT. Since the decision can involve the person with dementia, family member, and/or informal or professional carers, it is necessary to specify how the ethical issue was resolved. For example, in extreme cases, a decision may be made to use an AT without the knowledge of the person with dementia if they have an advanced level of cognitive impairment. For example, shoes with an implanted sensor may be used for tracking purposes. In such cases, the decision-makers should be aware of the consequences of using this technology and of the ethical and legal implications.

8. *Implement chosen solutions*: When the AT is implemented, it must be clear who is responsible for quality assurance, continuing maintenance and services, and continuing assessment of the efficiency and effectiveness of the AT. Maintaining use of the AT, especially when it involves 'work' by the person with dementia, is a critical issue, as use may be irregular, discontinuous or inadequate.

9. *Assess the results*: Regular, continuous monitoring of the AT's use is necessary not just to assess its effectiveness in meeting the needs of the person with dementia and solving the problem, but also to assess it with regard to ethical issues, usability, acceptability and ease of maintenance. At the same time, the status of the person with dementia must be continually assessed in terms of their cognitive impairment, behavioural issues, needs and problems in order to determine whether adjustments need to be made to the way the AT is used. Features of the person with dementia and the environment that influence their use of the AT should also be assessed so adjustments can be made to ensure the AT is effective for the individual in question. An assessment should also be made of whether the implementation and use of the AT follows the principles of good practice.

Appropriate, efficient and effective use of AT in persons with dementia can result when these steps are followed rigorously. AT is a tool that can improve the quality of life of people with dementia, but it should be used only if it enables them to be as independent as possible with a quality of life that encompasses various life-enhancing activities and leaves their integrity and personhood intact.

References

Alzheimer's Association (2016a) Alzheimer's disease facts and figures. Available via DIALOG. https://www.alz.org/documents_custom/2016-facts-and-figures.pdf. Accessed 06 Aug 2016

Alzheimer's Association (2016b) Stages of Alzheimer's. Available via DIALOG. https://www.alz.org/alzheimers_disease_stages_of_alzheimers.asp?type=carccenter_footer. Accessed 06 Aug 2016

Alzheimer Europe (2012) The ethical issues linked to the use of assistive technology in dementia care. Available via DIALOG. http://www.alzheimer-europe.org/Ethics/Ethical-issues-in-practice/The-ethical-issues-linked-to-the-use-of-assistive-technology-in-dementia-care. Accessed 05 Aug 2016

Alzheimers.net (2014) 10 Lifesaving location devices for dementia patients. Available via DIALOG. www.alzheimers.net/8-8-14-location-devices-dementia. Accessed 06 Aug 2016

Alzheimer's Society (2014) Assistive technology—Devices to help with everyday living. Available via DIALOG. http://www.alzheimers.org.uk/site/scripts/documents_info.php?documentID=109. Accessed 06 Aug 2016

Alzheimer's Society (2011) Alzheimer's Society Position Paper on Assistive Technology. Available via DIALOG. https://www.alzheimers.org.uk/site/scripts/download_info.php?fileID=1098. Accessed 06 Aug 2016

Alzheimer's Society of Canada (2011) Guidelines for care: person-centered care of people with dementia living in care homes. Framework. Available via DIALOG. http://www.alzheimer.ca/~/media/Files/national/Culture-change/culture_change_framework_e.pdf. Accessed 06 Aug 2016

AtDementia (2007) The ethical uses of assistive technology. Available via DALOG. https://www.atdementia.org.uk/content_files/files/The_ethical_use_of_assistive_technology.pdf. Accessed 06 Aug 2016

Bjørneby S, Topo P, Holthe T (1999) Technology, ethics and dementia: a guidebook on how to apply technology in dementia care. Norwegian Centre for Dementia Research, Oslo. Available via DIALOG. https://www.uni-bamberg.de/fileadmin/uni/fakultaeten/sowi_professuren/urbanistik/ted.pdf. Accessed 06 Aug 2016

Bonner S, Idris T (2012) Assistive Technology as a Means of supporting people with dementia: a review. Housing Learning & Improvement Network, London. Available via DIALOG. http://www.housinglin.org.uk/Topics/browse/HousingandDementia/Provision/AssistiveTechnology/?parent=5052&child=8563. Accessed 08 Aug 2016

Gibson G, Dickinson C, Brittain K et al. (2015) The everyday use of assistive technology by people with dementia and their family carers: a qualitative study. BMC Geriatr 15(89). doi:10.1186/s12877-015-0091-3

Gibson G, Newton L, Pritchard G et al (2016) The provision of assistive technology products and services for people with dementia in the United Kingdom. Dementia 15(4):681–701. doi:10.1177/1471301214532643

Godwin B (2012) The ethical evaluation of assistive technology for practitioners: a checklist arising from a participatory study with people with dementia, family and professionals. J Assist Technol 6:123–135. doi:10.1108/17549451211234975

Kosta E, Pitkänen O, Niemelä M et al (2010) Mobile-centric ambient intelligence in health- and homecare—anticipating ethical and legal challenges. Sci Eng Ethics 16(2):303–323. doi:10.1007/s11948-009-9150-5

Landau R, Auslander GK, Werner S et al (2011) Who should make the decision on the use of GPS for people with dementia? Aging Ment Health 15(1):78–84. doi:10.1080/13607861003713166

Landau R, Werner S, Auslander G et al (2010) What do cognitively intact older people think about the use of electronic tracking devices for people with dementia? A preliminary analysis. Int Psychogeriatr 22(8):1301–1309. doi:10.1017/S1041610210001316

Lindqvist E, Larsson TJ, Borell L (2015) Experienced usability of assistive technology for cognitive support with respect to user goals. NeuroRehabilitation 36(1):135–149. doi:10.3233/NRE-141201

Lindqvist E, Nygård L, Borell L (2013) Significant junctures on the way towards becoming a user of assistive technology in Alzheimer's disease. Scand J Occup Ther 20(5):386–396. doi:10.3109/11038128.2013.766761

Magnusson L, Hanson EJ (2003) Ethical issues arising from a research, technology and development project to support frail older people and their family carers at home. Health Soc Care Community 11(5):431–439

Marshall M (ed) (2000) Astrid: a social and technological response to meeting the needs of individuals with Dementia and their carers. Hawker Publications, London

Nuffield Council on Bioethics (2009) Dementia: ethical issues. Nuffield Council on Bioethics, London

O'Keefe J, Maier J, Freiman MP (2010) Assistive technology for people with dementia and their caregivers at home: what might help. Administration on Aging, Washington, DC. Available via DIALOG. https://www.google.com/#q=Assistive+technology+for+people+with+dementia+and+their+caregivers+at+home:+What+might+help. Accessed 08 Aug 2016

Pai MC, Lee CC (2016) The incidence and recurrence of getting lost in community-dwelling people with Alzheimer's Disease: a two and a half-year follow-up. PLoS ONE 11(5):e0155480. doi:10.1371/journal.pone.0155480

Rialle V, Ollivet C, Guigui C et al (2008) What do family caregivers of Alzheimer's disease patients' desire in smart home technologies? Contrasted results of a wide survey. Methods Inf Med 47(1):63–69

Robinson L, Brittain K, Stephen L et al (2009) Keeping in touch everyday (KITE) project: developing assistive technologies with people with dementia and their carers to promote independence. Int Psychogeriatr 21(3):494–502. doi:10.1017/S1041610209008448

Rosenberg L, Kottorp A, Nygard L (2012) Readiness for technology use with people with dementia: the perspectives of significant others. J Appl Gerontol 31(4):510–530. doi:10.1177/0733464810396873

Smebye KL, Kirkevold M, Engedal K (2016) Ethical dilemmas concerning autonomy when persons with dementia wish to live at home: a qualitative, hermeneutic study. BMC Health Serv Res 16(21). doi:10.1186/s12913-015-1217-1

Woolham J (ed) (2006) Assistive technology in dementia care: developing the role of technology in the care and rehabilitation of people with dementia—current trends and perspectives. Hawker Publications, London

Zwijsen SA, Niemeijer AR, Hertogh CM (2011) Ethics of using assistive technology in the care for community-dwelling elderly people: an overview of the literature. Aging Ment Health 15(4):419–427. doi:10.1080/13607863.2010.543662

Erratum to: Living Safely and Actively in and Around the Home: Four Applied Examples from Avatars and Ambient Cubes to Active Walkers

Martin Biallas, Edith Birrer, Daniel Bolliger, Andreas Rumsch, Rolf Kistler, Alexander Klapproth and Aliaksei Andrushevich

Erratum to:
Chapter 2 in: I. Kollak (ed.), *Safe at Home with Assistive Technology*,
DOI 10.1007/978-3-319-42890-1_2

In the original version of the book, Chapter 2 was inadvertently published without the co-author name "Aliaksei Andrushevich" and included his biography in Back matter "CVs of Authors". The erratum chapter and the book have been updated with the change.

The updated original online version for this book can be found at 10.1007/978-3-319-42890-1_2

M. Biallas · E. Birrer · D. Bolliger · A. Rumsch · R. Kistler (✉) · A. Klapproth ·
A. Andrushevich
Engineering and Architecture (CH), iHomeLab,
Lucerne University of Applied Sciences and Arts, Luzern, Switzerland
e-mail: rolf.kistler@hslu.ch

A. Andrushevich
e-mail: aliaksei.andrushevich@hslu.ch

© The Author(s) 2017   E1
I. Kollak (ed.), *Safe at Home with Assistive Technology*,
DOI 10.1007/978-3-319-42890-1_12

CVs of Authors

Aliaksei Andrushevich

He is a senior researcher at the iHomeLab of the Lucerne University of Applied Sciences and Arts. His research focuses on the Internet of Things, Intelligent Living, Assistive Systems, Energy Efficiency and Sustainability issues within home, Embedded Systems, Software Development and Algorithms for Building Intelligence. He is also interested in the effects of information technology on the economy and society. Aliaksei Andrushevich graduated as a mathematician and software developer from Belarusian State University in Minsk in 2006 and possesses MSc in Embedded Systems Design from the University of Lugano in 2007. He actively takes on tasks as a reviewer and as a member of several technical program committees. Aliaksei Andrushevich is a member of IEEE, ACM, and also of AAATE. He has been recognized with the AAL Forum Young Researcher Award in 2014. Since 2015 he is also acting as a guest lecturer for the Internet of Things at the Hogeschool van Amsterdam.

Alexander Bejan

M.Sc., academic researcher, Faculty Health, Safety, Society, Furtwangen University of Applied Sciences (D)
Book contribution
Evaluation and outcomes of assistive technologies in an outpatient setting—a technical-nursing science approach
Fields of interest
Human-technology interaction research and development, ambient assisted living (esp. in smart home contexts), technology-supported (collaborative) care
Publications (selection)
Bejan A, Lindwedel-Reime U (in print) Moderne computer basierte assistive Technologien für kranke Kinder und Jugendliche: Ein kurzer Überblick über den Stand der Technik. JuKiP Fachmagazin für Gesundheits- und Kinderkrankenpflege
Bejan A (in print) Multimediale Biografiearbeit: Wie können assistive Technologien Menschen mit Demenz helfen, sich an die eigene Vergangenheit zu crinnern? Journal dess_orientiert

© The Author(s) 2017
I. Kollak (ed.), *Safe at Home with Assistive Technology*,
DOI 10.1007/978-3-319-42890-1

Further information and contact
http://mensch-technik-teilhabe.de/team/alexander-bejan/
beja@hs-furtwangen.de

Martin Biallas

Ph.D. in biomedical engineering from the Swiss Federal Institute of Technology
(ETH) and a degree in electrical engineering and information technologies from the
University of Karlsruhe, senior research associate in the ambient assisted living
(AAL) team, iHomeLab, Lucerne University of Applied Sciences and Arts—
Engineering and Architecture (CH)
Book contribution
Living safely and actively in and around the home—Four applied examples from
avatars and ambient cubes to active walkers
Fields of interest
   Assistive communication and safety systems for people living at home, stress
measurement and intervention mechanisms, connected hearing aids, wearables,
augmented reality and the Internet of Things
Recent work and publications
https://www.hslu.ch/de-ch/technik-architektur/forschung/kompetenzzentren/ihome-
lab/projekte/
Further information and contact
www.iHomeLab.ch
martin.biallas@hslu.ch

Edith Birrer

Degree in computer sciences from the Swiss Federal Institute of Technology
(ETH), researcher, iHomeLab, Lucerne University of Applied Sciences and Arts—
Engineering and Architecture (CH)
Book contribution
Living safely and actively in and around the home—Four applied examples from
avatars and ambient cubes to active walkers
Fields of interest
Mobile applications and reusable frameworks
Recent work and publications
https://www.hslu.ch/de-ch/technik-architektur/forschung/kompetenzzentren/ihome-
lab/projekte/
Further information and contact
www.iHomeLab.ch
edith.birrer@hslu.ch

Daniel Bolliger

Ph.D. in natural sciences and a degree in experimental physics from the Swiss
Federal Institute of Technology (ETH), senior research associate, iHomeLab,

Lucerne University of Applied Sciences and Arts—Engineering and Architecture (CH)
Book contribution
Living safely and actively in and around the home—Four applied examples from avatars and ambient cubes to active walkers
Fields of interest
Energy efficiency and ambient assisted living
Recent work and publications
https://www.hslu.ch/de-ch/technik-architektur/forschung/kompetenzzentren/ihome-lab/projekte/
Further information and contact
www.iHomeLab.ch
Daniel.bolliger@hslu.ch

Jane Burridge

Ph.D., graduated physiotherapist, trained musician playing and teaching the flute, professor of restorative neuroscience, leader of the neurorehabilitation research group, University of Southampton (UK)
Book contribution
Arm rehabilitation at home for people with stroke: staying safe—Encouraging results from the co-designed LifeCIT programme
Fields of interest
Improving recovery of movement following acquired brain damage, understanding the mechanisms associated with normal, loss and recovery of motor function, using the Internet to support home-based rehabilitation, crossing traditional rehabilitation boundaries, collaborating with engineers, neuroscientists and psychologists and particularly understanding how rehabilitation technologies can translate into clinical practice
Recent grants
Wearable sensors to support home-based rehabilitation and generate objective measures of impairment (NIHR i4i)
M-MARK (2015–2017) Mechanical Muscle Activity with Real-time Kinematics: A novel combination of existing technologies to improve arm recovery following stroke (NIHR i4i)
INSPIRE (2013–2016) Feasibility study of an iCycle for functional recovery after incomplete spinal cord injury with externally funded Ph.D. studentship (PI: INSPIRE)
Publications (selection)
Wee SK, Hughes AM, JH Burridge JH et al. (2015) Effect of trunk support on upper extremity Function in people with chronic stroke and healthy controls. Phys Ther 95(8):1163–1171 doi:10.2522/ptj.20140487
Tedesco Triccas L, Burridge JH, Hughes AM et al (2015) Multiple sessions of transcranial direct current stimulation and upper extremity rehabilitation in stroke: A review and meta-analysis. Clin Neurophysiol 127(1):946–955. doi:10.1016/j.clinph.2015.04.067

Wee SK, Hughes AM, Burridge JH et al. (2014) Trunk restraint to promote upper extremity recovery in stroke patients: a systematic review and meta-analysis. Neurorehabil Neural Repair 28(7):660–677. doi:10.1177/1545968314521011

Johnson L, Burridge JH, Demain SH (2013) Internal and external focus of attention during gait re-education: an observational study of physical therapist practice in stroke rehabilitation. Phys Ther 93(7):957–966. doi:10.2522/ptj.20120300

Hughes AM, Burridge JH, Denain SH et al. (2014) Translation of evidence-based Assistive Technologies into stroke rehabilitation: users'perceptions of the barriers and opportunities.

BMC Health Serv Res.14:124. doi:10.1186/1472-6963-14-124

Website: http://www.southampton.ac.uk/healthsciences/about/staff/jane_burridge.page

Further information and contact

J.H.Burridge@soton.ac.uk

Anna Lena Grans

M.A., research assistant RetroBrain research and development UG and Humboldt University of Berlin (D)

Book contribution

Using gaze control for communication and environment control—How to find a good position and start working

Fields of interest

Augmentative and alternative communication, assistive technologies for communication, technology and dementia studies

Recent work

Evaluation of preventive and health-promoting aspects of the use of technical games in dementia. Research projects on gaze control for communication aids and on qualification of staff for using assistive technologies

Publications (selection)

Saborowski M, Grans AL, Kollak I (2015) Wenn Blicke die Kommunikation steuern: Beobachtung einer Augensteuerung im Alltag. In: Antener G, Blechschmidt A, Ling K (eds) UK wird erwachsen. Initiativen in der Unterstützten Kommunikation. Von Loeper, Karlsruhe, pp 370–383

Knorz G, Saborowski M, Grans LA et al. (2015) „Entschleunigte Kommunikation oder beschleunigtes Missverstehen?" (2015) In: *alice—das Hochschulmagazin der ASH Berlin* 29:57–59

Saborowski M, Grans AL, Thedinga M et al. (2016) „Dann würde ich versuchen, eine Augensteuerung zu beantragen." Befragung von Fachpersonen aus der Unterstützten Kommunikation zu Erfahrungen mit Augensteuerungen. Zeitschrift für Heilpädagogik 1:16–28

Grans AL, Saborowski M, Kollak I (2016) Sprechen, Lachen, Rückwärtsfahren—was Kommunikationshilfen im Alltag erleben. Unterstützte Kommunikation 2:44–46

Further information and contact
http://www.eyetrack4all.de
https://memore.de/
anna.lena.grans@hu-berlin.de

Josef M. Huber

M.A. nursing science, B.A. nursing management, doctoral student, scientific advisor in research and development, Evangelische Heimstiftung GmbH, Centre for innovation, Stuttgart (D)

Book contribution

Use and development of new technologies in public welfare services—A user-centred approach using step by step communication for problem solving

Fields of interest

Citizen science, knowledge management, voluntary work

Recent research

QuartrBack (2015–2018) http://www.quartrback.de (Federal Ministry of Education and Research) (ubcontractor)

AICASy (2015–2017) http://www.mtidw.de/ueberblick-bekanntmachungen/ALS/aicasys) (subcontractor)

MOVEMENZ (2015–2016) https://www.itas.kit.edu/projekte_deck14_movemenz.php. FUgE (2013–2014) http://ankom.his.de/projekte/p13_hs_esslingen

Publications (selection)

Huber JM, Eckstein C, Riedel A et al. (2016) Bildungsübergänge durch Tutorien erfolgreich gestalten. Den Aufbau von Handlungskompetenz prinzipienorientiert und durch Reflexion in Peergruppen begleiten. In: PADUA Fachzeitschrift für Pflegepädagogik, Patienteneducation und -bildung 11(1):45–51

Linde AC, Huber JM, Riedel A (2015) Ansatzpunkte und Empfehlungen zur Stärkung ethischer Reflexion und Ethikkompetenz in der (Pflege-)Praxis. Paper presented 3-Länder-Konferenz, Konstanz, 21 September 2015

Kimmerle B, Huber JM, Riedel A et al. (2015) Pflegeberuflich Qualifizierte: Betrachtung einer Studierendengruppe beim Übergang in die Hochschule. In: Freitag WK, Buhr R, Danzeglocke EM et al. (eds) Übergänge gestalten. Durchlässigkeit zwischen beruflicher und hochschulischer Bildung erhöhen. Waxmann Verlag, Bonn, pp 151–172

Riedel A, Kimmerle B, Bonse-Rohmann M et al. (2015) Spannungsfelder am Übergang von der beruflichen Bildung und Praxis an die Hochschule. Pädagogik der Gesundheitsberufe 2(1):25–35

Riedel A, Huber JM, Linde AC (2013) Wiederkehrende ethische Dilemmata strukturiert reflektieren—Psychiatrische Pflegepraxis. CNE Schwerpunkt Beratung. PsychPflege 19:261–268

Further information and contact

www.ev-heimstiftung.de

j.huber@ev-heimstiftung.de

Ann-Marie Hughes

Ph.D. in electronics and electrical engineering, chartered physiotherapist, associate professor in rehabilitation technologies at the faculty of health sciences, University of Southampton (UK)

Book contribution
Arm rehabilitation at home for people with stroke: staying safe—Encouraging results from the co-designed LifeCIT programme

Fields of work
User needs, novel rehabilitation technologies including robotics, functional electrical stimulation, internet-based motivational rehabilitation and non-invasive brain stimulation, multidisciplinary, research-led education and stakeholder engagement

Recent research
MRC (2016–2018) Low-cost personalised instrumented clothing with integrated FES electrodes for upper limb rehabilitation (co-investigator)
Public Engagement with Research (2015–2016) University of Southampton (co-investigator)

Publications (selection)
Hughes AM, Freeman C, Banks T et al. (2016) Using the Tuning Methodology to design the founding benchmark competences for a new academic professional field: the case of advanced rehabilitation technologies. *Tuning Journal for Higher Education*:1–26
Tedesco Triccas L, Burridge JH, Hughes AM et al. (2015) A double-blinded randomised controlled trial exploring the effect of anodal transcranial direct current stimulation and uni-lateral robot therapy for the impaired upper limb in sub-acute and chronic stroke. *Neurorehabilitation*:1–28
Tedesco Triccas L, Burridge JH, Hughes AM et al. (2015) Multiple sessions of transcranial direct current stimulation and upper extremity rehabilitation in stroke: a review and meta-analysis. *Clinical Neurophysiology*: 1–40
Wee SK, Hughes AM, Warner MB et al. (2015) Effect of trunk support on upper extremity function in people with chronic stroke and people who are healthy. *Physical Therapy* 95(8):1163–1171
Hughes AM, Hallewell E, Kutlu M et al. (2014) Combining electrical stimulation mediated by iterative learning control with movement practice using real objects and simulated tasks for post-stroke upper extremity rehabilitation. *Neurosonology and Cerebral Hemodynamics* 10(2):117–122

Further information and contact
http://www.southampton.ac.uk/healthsciences/about/staff/ah10.page
A.Hughes@soton.ac.uk

Natalie Jankowski

M.A., research assistant, Institut für Rehabilitationswissenschaften, Abt. Rehabilitationstechnik, Neue Medien, Humboldt University of Berlin (D)
Book contribution
Telemonitoring in home care –Creating the potential for a safer life at home

Fields of interest
Telemedizin, Altern und Technik, Neue Technologien in der Rehabilitation, Usabillity Forschung und Nutzerpartizipation, Science and Technology Studies (STS)

Publications (selection)
Jankowski N, Gerstmann J, Wahl M (2016) Nutzungsbereitschaft von Telemedizin in der Schlaganfallnachsorge—Sicht der Behandler. In: Schug S, Schmücker P, Semler S et al. (eds) E-Health-Rahmenbedingungen im europäischen Vergleich: Strategien, Gesetzgebung, Umsetzung. AKA Verlag, Berlin, pp 133–142

Jankowski N, Gerstmann J, Wahl M (2015) Unterstützungsansätze in der Versorgung mittels mobiler Geräte und Telemedizin. In: Weisbecker A, Burmester M, Schmidt A (eds) Mensch und Computer 2015—Workshopband. De Gruyter, Berlin, Boston, pp 117–124

Jankowski N et al. (2014) Technologies in Practice. In: Bister M, Niewöhner J (eds) Alltag in der Psychiatrie im Wandel. Ethnographische Perspektiven auf Wissen, Technologie und Autonomie. Panama Verlag, Berlin

Megges H, Jankowski N, Peters O (2013) Caregiver needs analysis for product development of an assistive technology system in dementia care. In 23th Alzheimer Europe Conference, St. Julian's, Malta

Kruse A, Schmitt E, Holfelder JD et al. (2011) Kreativität im Alter—Ergebnisse der Auswertung von Bewerbungen zum Otto-Mühlschlegel-Preis. In: Kruse A (ed) Kreativität im Alter. Universitätsverlag Winter, Heidelberg, pp 195–234

Further information and contact
https://www.reha.hu-berlin.de/personal/mitarbeiter/1688048
jankowna@hu-berlin.de

Katja Joronen

Ph.D, RN, adjunct professor, Tampere University Hospital, Science center and School of Health Sciences, University of Tampere (FI)

Book contribution
Parents' experiences of caring for a ventilator-dependent child: a review of the literature

Fields of interest
Child and adolescent health, subjective well-being and pediatric nursing, new methods for health promotion of children and their families (e.g. mobile games), secondary data analysis (e.g. school health promotion study)

Publications (selection)
Räsänen T, Lintonen T, Joronen K et al. (2015) Girls and boys gambling with health and well-being. Journal of School Health 85(4):214–222

Konu A, Joronen K, Lintonen T (2014) Seasonality in School Well-being: The Case of Finland. Child Indicators Research 8(2):265–277. doi:10.1007/s12187-014-9243-9

Joronen K, Konu A, Rankin SH et al. (2012) An evaluation of a drama program to enhance social relationships and anti-bullying at elementary school: a controlled study. Health Promotion International 27(1):5–14. doi:10.1093/heapro/dar012

Lepistö S, Joronen K, Åstedt-Kurki P et al. (2012) Subjective well-being in Finnish adolescents experiencing family violence. Journal of Family Nursing 18(2):200–233. doi:10.1177/1074840711435171
Joronen K, Häkämies A, Åstedt-Kurki P (2011) Children's experiences of drama programme of social and emotional learning. Scandinavian Journal of Caring Sciences (25):671–678. doi:10.1111/j.1471-6712.2011.00877.x
Further information and contact
http://www.uta.fi/hes/yhteystiedot/henkilosto/joronen.html
katja.joronen@uta.fi

Anke S. Kampmeier

Prof. Dr., professor of social work and work with people with disabilities at the department of social work and education, University of Applied Sciences Neubrandenburg (D)
Book contribution
Caring TV—for older people with multimorbidity living alone—Positive feedback from users in Berlin and rural Mecklenburg-West Pomerania
Fields of interest
Inclusion, transdisciplinarity, professionalism and cooperation for assisting disadvantaged persons in the vocational training system
Recent research
INKLUSIV (2016–2019) Education in Mecklenburg-West Pomerania (Ministry of Education and Research)
SaLSA (2011–2014) Self-determined and Activated Life and Security in Old Age—a pilot study on the integration of Caring TV into the elderly support sector (Federal Ministry of Education and Research)
Personal Budget (2009–2010) (Excellency Program of Mecklenburg-West Pomerania)
Publications (selection)
Kampmeier AS, Kraehmer S, Schmidt S (Eds) (2014) Das Persönliche Budget. Selbständige Lebensführung von Menschen mit Behinderungen. Kohlhammer Verlag, Stuttgart
Kampmeier AS, Kraehmer K (2014) Inklusion als Chance für Hochschulen und Hochschuldidaktik. Die Neue Hochschule. DNH. Heft 4—Jahrgang 2014:110–113
Kampmeier AS (2010) Transition zwischen den Paradigmen—Stolperstein—Persönliches Budget. In: Schildmann U (ed) Umgang mit Verschiedenheit in der Lebensspanne. Klinkhardt Verlag, Bad Heilbrunn
Kampmeier AS (2010) Realisierung des Persönlichen Budgets in der Hilfe für Menschen mit Behinderungen. In: Michel-Schwartze B (ed) "Modernisierungen" methodischen Handelns in der Sozialen Arbeit. VS Verlag, Wiesbaden
www.hs-nb.de
kampmeier@hs-nb.de

Hesook Suzie Kim

Ph.D. Sociology (Brown University), M.A. Sociology (Brown University), M.S. Nursing Education (Insiana University), B.S. Nursing (Indiana University), professor II, Institutt for helsefag, University College of Southeast Norway, Drammen, Norway, professor emerita, College of Nursing, University of Rhode Island, Kingston, RI (USA)

Book contribution

Assistive technology for people with dementia: ethical considerations

Recent Research

Crisis resolution and home treatment in community mental health service: development, practice, experiences, and outcomes (2008–2011) Buskerud University College, Institute of Health Sciences (funded by Norwegian Research Council) (project director)

Post-stroke fatigue: characteristics, related experiences and pathophysiological mechanisms in a longitudinal perspective (2007–2010) Buskerud University College, Institute of Health Sciences (funded by Norwegian Research Council) (project director)

Publications (selection)

Kim HS (2015) The essence of nursing practice. Springer Publishing, New York

Kim HS, Schwartz-Barcott D, Holter IM (2012) Cross-cultural use and validity of pain scales and questionnaires—Norwegian case study. In: Incayawar M, Todd KH (eds) Pain, Culture, Brain and Analgesia—Understanding and managing pain in diverse populations. Oxford University Press, New York

Kim HS (2012) Action science. In: Fitzpatrick J (ed) Encyclopedia of nursing research. Springer Publications, New York

Kim HS, Clabo LML, Burbank P et al. (2010) Application of critical reflective inquiry in nursing education. In: Lyons N (ed) *Handbook of reflection and reflective inquiry*. Springer Science+Business Media, New York, pp 159–172

Kim HS (2010) The nature of theoretical thinking in nursing, 3rd ed. Springer Publishing, New York

Further information and contact

hsuziekim@comcast.net

Beatrix Kirchhofer

M.Sc., academic researcher, faculty health, safety, society, Furtwangen University of Applied Sciences (D)

Book contribution

Evaluation and outcomes of assistive technologies in an outpatient setting—a technical-nursing science approach

Fields of interest

Nursing and health sciences, ambient assisted living (AAL)

Further information and contact
http://mensch-technik-teilhabe.de/team/beatrix-kirchhofer/
kix@hs-furtwangen.de

Rolf Kistler

Degree in electrical engineering (industrial informatics), leader of the AAL research group at the iHomeLab, Lucerne University of Applied Sciences and Arts—Engineering and Architecture (CH)
Book contribution
Living safely and actively in and around the home—Four applied examples from avatars and ambient cubes to active walkers
Fields of interest
Building intelligence and smart homes with focus on ambient assisted living, socio-technical systems, geronto-technology, human building interaction, activity recognition, Internet of Things, cyber-physical systems, energy efficiency, unified communication networks, wireless sensor networks and plug and play protocols at the iHomeLab
Recent work and publications
https://www.hslu.ch/de-ch/technik-architektur/forschung/kompetenzzentren/ihomelab/projekte/
Further information and contact
www.iHomeLab.ch
rolf.kistler@hslu.ch

Alexander Klapproth

Degree in engineering from the Swiss Federal Institute of Technology in Zurich (ETHZ), Professor, co-founder of the iHomeLab at Lucerne University of Applied Sciences and Arts—Engineering and Architecture (CH)
Book contribution
Living safely and actively in and around the home—Four applied examples from avatars and ambient cubes to active walkers
Recent research grants
HANNA—personal assistance system for people with dementia and their families
iBigCare—Investigation on the efficacy and applicability of assistive monitoring systems
Kith & Kin—Communication device (for the elderly) with simple options
My Life, My Way—a modern version of a butler
WizEE self-learning home energy management
Recent work and publications
https://www.hslu.ch/de-ch/technik-architektur/forschung/kompetenzzentren/ihomelab/projekte/
Further information and contact
www.iHomeLab.ch
alexander.klapproth@hslu.ch

Peter König

Prof. Dr., professor for nursing and rehabilitation management, faculty health, safety, society, University of Applied Sciences Furtwangen (D)

Book contribution
Evaluation and outcomes of assistive technologies in an outpatient setting—a technical-nursing science approach

Fields of interest
Evidence based Nursing, nursing concepts in dementia, assistive technologies in disabilities and care dependency classification and terminology

Recent research grants
Effekte eines pflegerischen Beratungs- und Anleitungsprogramms zur Prophylaxe von oraler Mukositis bei der Therapie mit 5-FU-haltigen Chemotherapeutika bei Patienten mit soliden Tumoren (Fördergesellschaft Forschung Klinik für Tumorbiologie)

ENAS: Effekte und Nutzen altersgerechter Assistenzsysteme—praktikable Vorgehensmodelle, Evaluationsmethoden und Werkzeuge (Bundesministerium für Bildung und Forschung)

Blickwechsel: Leben im Alter: Zukunftsperspektiven aus Geriatrie, Gerontologie, Pflege und Prävention (Robert-Bosch Stiftung). Besser Leben im Alter durch Technik (Bundesministerium für Bildung und Forschung)

Komfortküchen für ein selbstbestimmtes Leben im Alter (Der Kreis Anna Schaible Stiftung)

Publications (selection)
Müller-Staub M, Schalek K, König P (eds) (2016) Pflegeklassifikationen und Begriffssysteme. Hans Huber, Göttingen

König P (2016) Barrierefreiheit im öffentlichen Raum: Die Barrieresituation für Beeinträchtigte in Furtwangen. Vortrag Gesundheitskongress Furtwangen 16.3.2016

König P (2016) Inwieweit macht es Sinn, GCP bei klinischen Studien anzuwenden, die nicht unter das AMG oder MPG fallen?—Eine Beispielstudie. Vortrag 1. D|A|CH Symposium Klinische Prüfungen Freiburg 7.3.2016

König P (2015) Was wollen und brauchen wir wirklich an Telemedizin und eHealth in der Versorgung? Sicht der Pflege. Vortrag Symposium Telemedizin and Digitalisierung Gesundheitsbereich Stuttgart 14.10.2015

König P (2015) Development of the ICNP in the German-Speaking Countries. Vortrag International Council of Nurses Seoul Südkorea 20.6.2015

Further information and contact
http://mensch-technik-teilhabe.de/team/peter-koenig/
koep@hs-furtwangen.de

Ingrid Kollak

Prof. Dr. phil., professor, health and nursing management, management and quality development in the health sector at Alice Salomon University of Applied Sciences Berlin (D)

Book contribution
Editor of the book, Prerequisits—Assitive technologies between user centered assistance and,technicalization', Using gaze control for communication and environment control—How to find a good position and start working
Fields of interest
Psychosocial interventions for people with dementia, technical assistance for frail and old people, nursing science and theory, intercultural and international aspects of nursing and nursing education, self-care and yoga
Recent research grants
GLEPA (2015–2017) LGBTI* aging and care (Berlin Institute for Applied Research—IFAF Berlin) (co-director)
EyeTrack (2013–2016) Expansion of the user group for gaze controlled augmentative and alternative communication through the development of new processes of gaze control (German Ministry of Education and Research) (project director)
Nursing council in Berlin (2014–2015) Representative survey on the acceptance of a nursing council in the federal state of Berlin (Senate for health and welfare Berlin) (project director)
Tales+Dementia+Study (2014–2015) Evaluation of the project "Once upon a time... Fairy tales and dementia" (Federal Ministry of Family Affairs, Senior Citizens, Women and Youth) (project director)
MAAL (2011–2014) Development of a part-time interdisciplinary Master program in Ambient Assisted Living (German Ministry of Education and Research) (co-director)
Publications (selection)
Kollak I (ed) (2016) Menschen mit Demenz durch Kunst und Kreativität aktivieren. Springer Verlag, Berlin, Heidelberg
Kollak I, Schmidt S (2016) Instrumente des Care und Case Management Prozesses. Springer Verlag, Berlin, Heidelberg
Jaccarini GA, Kollak I, Schmdit S (2015) Just in Case: Care and Case Management in Malta. MMDNA
Kollak I, Schmidt S (2015) Fallübungen Care und Case Management. Springer Verlag, Berlin, Heidelberg
Kollak I (2014) Time Out—Übungen zur Selbstsorge und Entspannung für Gesundheitsberufe. Springer Verlag, Berlin, Heidelberg
Further information and contact
www.ash-berlin.eu/hsl/kollak
kollak@ash-berlin.eu

Minna Lahtinen

Master of Health Sciences, RN, Tampere University Hospital, Science center and School of Health Sciences, University of Tampere (FI)
Book contribution
Parents' experiences of caring for a ventilator-dependent child: a review of the literature

Fields of interest
Pediatric nursing, pediatric intensive care, secretary of the regional ethics committee of the expert responsibility area of Tampere University Hospital, research coordinator of the Finnish Clinical Biobank Tampere
Publications (selection)
Lahtinen M, Rantala A, Heino-Tolonen T et al. (2016) Lääkkeetöntä kivunlievitystä edistävät ja estävät tekijät lasten sairaalahoidon aikana. [Factors enhancing or hindering the use of non-pharmacological pain relief methods among hospitalized children.] Tutkiva Hoitotyö 14(2):4–13
Lahtinen M, Rantala A, Heino-Tolonen T et al. (2015) Lääkkeetön kivunlievitys ja sen kirjaaminen lasten sairaalahoidon aikana. [Non-pharmacological pain relief and its documentation among hospitalized children.] Hoitotiede 27(4):324–337
Lahtinen M, Joronen K (2014) Vanhempien kokemukset hengityslaitetta tarvitsevan lapsen hoitamisesta kotona—kirjallisuuskatsaus. [Parents' experiences of caring for ventilator-dependent child—a literature review.] Hoitotiede 26(2):89–100
Further information and contact
http://www.pshp.fi/fi-FI/Yhteystiedot/Henkilot/henkilokortti?us_userid=77982
minna.maa.lahtinen@pshp.fi

Ulrike Lindwedel-Reime

M.Sc., academic researcher, faculty health, safety, society, University of Applied Sciences Furtwangen (D)
Book contribution
Evaluation and outcomes of assistive technologies in an outpatient setting—a technical-nursing science approach
Fields of interest
Nursing and health sciences, assistive technologies in disabilities and care dependency
Publications (selection)
Lindwedel-Reime U, König P, Kunze C et al. (2015) Effekte und Nutzen altersgerechter Assistenzsysteme—praktikable Vorgehensmodelle, Evaluationsmethoden und Werkzeuge (ENAS). Vortrag 6. DGP-Hochschultag Freiburg
Bejan A, Lindwedel-Reime U (in print) Moderne computerbasierte assistive Technologien für kranke Kinder und Jugendliche: Ein kurzer Überblick über den Stand der Technik. JuKiP Fachmagazin für Gesundheits- und Kinderkrankenpflege
Further information and contact
http://mensch-technik-teilhabe.de/team/ulrike-lindwedel-reime/
liru@hs-furtwangen.de

Carmen Llorente Barroso

Dr., adjunct professor in the department of audiovisual communication and publicity at the Universidad CEU San Pablo in Madrid (E)

Book contribution
Empowering the elderly and promoting active ageing through the Internet—The benefit of e-inclusion programmes
Fields of interest
Rhetorical advertising strategies, communicative effectiveness indicators associated with consumer behavior, new media, vulnerable audiences and digital divide
Recent work
Comunicación digital en las instituciones sanitarias para un envejecimiento activo, reference USPBSPPC03/2012, directed by Karen Sanders (funded by the University CEU San Pablo)
Brecha digital y mayores (PROVULDIG-CM), reference H2015/HUM-3434, directed by Leopoldo Abad Alcalá /Ignacio Blanco (funded by the Dirección General de Universidades e Investigación de la Consejería de Educación, Juventud y Deporte (CAM)
Personas mayores, e-commerce y administracion electrónica: hacia la ruptura de la tercera brecha digital, reference CSO2015-66746-R, directed by Leopoldo Abad Alcalá (funded by the Convocatoria 2015—Proyectos I+D+I. Programa Estatal de Investigación, Desarrollo e Innovación (MINECO))
Publications (selection)
García-Guardia ML, Llorente-Barroso C (2016) An empirical study of the recreational-expressive and referential roles of the cell phone among young students and their potential applications to advertising. Prisma Social, Especial (1):261–278
Sanders K, Sánchez-Valle M, Viñaras M et al. (2015) Do we trust and are we empowered by "Dr. Google"? Older Spaniards' uses and views of digital healthcare communication. Public Relations Review 41(5):794–800. Elsevier, New York. doi:10.1016/j.pubrev.2015.06.015
Llorente-Barroso C, García-García F (2015) The rhetorical construction of corporate logos. Arte, Individuo y Sociedad 27(2):289–309. doi:10.5209/rev_ARIS.2015.v27.n2.44667.
Llorente-Barroso C, Viñaras-Abad M, Sánchez-Valle M (2015) Mayores e Internet: La Red como fuente de oportunidades para un envejecimiento active. [Internet and the Elderly: Enhancing Active Ageing.] In: Comunicar 45:29–36. doi:10.3916/C45-2015-03
Further information and contact
www.uspceu@ceu.com
carmen.llorentebarroso@ceu.es

Claire Meagher

M.Sc. Physiotherapy, B.Sc. Psychology, Research fellow within health sciences, University of Southampton (UK)
Book contribution
Arm rehabilitation at home for people with stroke: staying safe—Encouraging results from the co-designed LifeCIT programme

Fields of interest
Cross-disciplinary research embedded in 21st century rehabilitation, with a focus on the increased use of technology to support rehabilitation and self-management for stroke survivors. Collaboration with interface design and signal processing engineers, psychologists and industrial partners. Development of new technologies for neurological rehabilitation.

Recent grants
Wearable sensors to support home-based rehabilitation and to generate objective measures of impairment (NIHR i4i) in collaboration with Imperial College London, industrial product design partners and NHS rehabilitation centers.

Publications (selection)
Meagher C, Burridge J, Ewings S et al. (2016) Feasibility Randomised Control Trial of LifeCIT, a web-based support programme for Constraint Induced Therapy (CIT) following stroke compared with usual care: Submitted. Physical Therapy: 1–24

Meagher C, Demain S, Anderson L et al. (2015) Understanding the burden of care during early supported discharge on spouses of people with stroke. At European Stroke Organisation Conference, Glasgow, UK 17–19th Apr 2015

Meagher C, Demain S (2015) A qualitative investigation into the burden of care during early supported discharge on spouses of people with stroke. At European Stroke Organisation Conference, Glasgow, UK 17–19th Apr 2015

Burridge J, Meagher C, Hughes AM et al. (2012) Development and pilot evaluation of a web-supported programme of Constraint Induced Therapy following stroke (LifeCIT). South West Stroke Network, Exeter, UK, 29th Apr 2015

Hughes AM, Freeman C, Meagher C et al. (2016) Innovative rehabilitation technologies for the home. In: World Federation for NeuroRehabilitation (WFNR) (symposium submitted)

Further information and contact
C.Meagher@soton.ac.uk

Claudia Nuß

B.A. research assistant, InterAktiv e.V Berlin (D)

Book contribution
Using gaze control for communication and environment control—How to find a good position and start working

Fields of interest
Augmentative and alternative communication, inclusive education

Recent work
Research project on gaze control for communication aids and on qualification of staff for using assistive technologies

Publications (selection)
Kollak I, Nuß C, Saborowski M (2016) Handreichung Augensteuerung: Hilfestellung für Vorüberlegungen, Planung und Einsatz einer Augensteuerung in

der Unterstützten Kommunikation. Available via DIALOG. https://nbn-resolving.
org/urn:nbn:de:kobv:b1533-opus-1334
Saborowski M, Nuß C, Kollak I (2016) Kommunikationshilfsmittel mit
Augensteuerung für nichtsprechende Personen mit schweren motorischen
Einschränkungen. In: Abstracts der Vorträge des 6. Heilberufe Science
Symposiums. HeilberufeSCIENCE Suppl 7:2–11, pp 4–5. doi:10.1007/s16024-
016-0270-y
Saborowski M, Nuß C, Kollak I (2015) „Taking a closer look at user-technology
relationships: a network model". In: Gross M, Klinski S von (eds): Research Day
„Stadt der Zukunft", Tagungsband der Beuth Hochschule Berlin. Mensch und Buch
Verlag, Berlin, pp 125–129
Nuß C, Saborowski M (2014) Neue Hilfsmittel in der Unterstützten
Kommunikation. Das Forschungsprojekt EyeTrack4all stellt sich vor. *alice*—das
Hochschulmagazin der ASH Berlin (27):38
Saborowski M, Nuß C, Kollak I (2014) How to conceptualise AAC
user-technology relationships: a study on eye control. Poster presentation (No 429)
ISAAC-Konferenz, Lissabon, Portugal 22.07.2014
Further information and contact
http://www.interaktiv-berlin.de/
http://www.eyetrack4all.de
c.nuss@interaktiv-berlin.de

Verena Pfister

M.Sc., Referentin für Altenhilfe, BruderhausDiakonie, Reutlingen (D)
Book contribution
Use and development of new technologies in public welfare services—A
user-centred approach using step by step communication for problem solving
Fields of interest
Durchführung von AAL-Projekten, Einbindung von Senioren-Experten,
Konzeptentwicklung für neuen Dienstleistungen mit Technikunterstützung
Recent work
KoBial (2016–2018) Kooperative Bauvorhaben im Sozialwesen—
Wertschöpfungssysteme und Service- Engineering (Ministry of Finance and
Economy Baden-Württemberg)
Assistent (2016–2018) Assistierte Kommunikation in ambulanten
Betreuungsformen, Innovationsprogramm Pflege (Ministry of Social Affaires and
Integration Baden-Württemberg)
EmAsIn (2015–2018) Emotionssensitive Assistenzsysteme zur reaktiven psychol-
ogischen Interaktion mit Menschen (German Ministry of Education and Research)
PflegeCoDe (2015–2018) Pflegecoaching für die optimale Unterstützung von
Menschen mit Demenz (German Ministry of Education and Research). NiviL

(2014–2017) nichtvisuelle Wirkung von Licht (German Ministry of Education and Research)
Publications (selection)
Pfister V, Steiner B (2015) Durch Interdisziplinarität zur AAL-Lösung für Senioren? Ergebnisse einer Bedarfsanalyse für eine individualisierte Assistenzplattform. VDE Verlag, Berlin
Denecke J, Felix F, Pfister V et al. (2016) Bedarfsanalysator zur Bestimmung eines Assistenzsystems (PATRONUS). VDE Verlag, Berlin
Further information and contact
www.bruderhausdiakonie.de
Verena.pfisterbruderhausdiakonie.de

Andreas Rumsch

Degree in electrical engineering from the Swiss Federal Institute of Technology (ETH) and a degree in economics for engineers, senior research associate in the smart energy team, iHomeLab, lecturer in the subject economies for engineers at the Lucerne University of Applied Sciences and Arts (CH)
Book contribution
Living safely and actively in and around the home—Four applied examples from avatars and ambient cubes to active walkers
Fields of interest
Mobility for impaired persons, load disaggregation, artificial intelligence, load management and Internet of Things
Recent work and publications
https://www.hslu.ch/de-ch/technik-architektur/forschung/kompetenzzentren/ihomelab/projekte/
Further information and contact
www.iHomeLab.ch
andreas.rumsch@hslu.ch

Maxine Saborowski

Dr., research assistant, Technical University Berlin (D)
Book contribution
Using gaze control for communication and environment control—How to find a good position and start working
Fields of interest
Science and technology studies, assistive technologies for communication, childhood studies *Recent work*
(Re)Organisation of professionalism and gender in social work
Research project on gaze control for communication aids and on qualification of staff for using assistive technologies

Publications (selection)
Kollak I, Nuß C, Saborowski M (2016) Handreichung Augensteuerung: Hilfestellung für Vorüberlegungen, Planung und Einsatz einer Augensteuerung in der Unterstützten Kommunikation. Available via DIALOG. https://nbn-resolving. org/urn:nbn:de:kobv:b1533-opus-1334
Saborowski M, Nuß C, Kollak I (2016) Kommunikationshilfsmittel mit Augensteuerung für nichtsprechende Personen mit schweren motorischen Einschränkungen. In: Abstracts der Vorträge des 6. Heilberufe Science Symposiums. HeilberufeSCIENCE Suppl 7:2–11, pp 4–5. doi:10.1007/s16024-016-0270-y
Saborowski M, Kollak I (2015) How do you care for technology?—Care professionals' experiences with assistive technology in care of the elderly. In: Technological Forecasting & Social Change (93): 133–140 Elsevier, Amsterdam. doi:10.1016/j.techfore.2014.05.006
Saborowski M, Grans AL, Kollak I (2015) Wenn Blicke die Kommunikation steuern: Beobachtung einer Augensteuerung im Alltag. In: Antener G, Blechschmidt A, Ling K (eds) UK wird erwachsen. Initiativen in der Unterstützten Kommunikation. Von Loeper, Karlsruhe, pp 370–383
Saborowski M, Nuß C, Kollak I (2015) Taking a closer look at user-technology relationships: a network model". In: Gross M, Klinski S von (eds): Research Day „Stadt der Zukunft", Tagungsband der Beuth Hochschule Berlin. Mensch und Buch Verlag, Berlin, pp 125–129
Further information and contact
http://www.eyetrack4all.de
http://www.ash-berlin.eu/organisation/wissenschaftliche-mitarbeiterinnen/dr-maxine-saborowski/
m.saborowski@tu-berlin.de

María Sánchez-Valle

Dr., Adjunct Professor in the department of audiovisual communication and publicity at
the Universidad CEU San Pablo in Madrid (E)
Book contribution
Empowering the elderly and promoting active ageing through the Internet—The benefit of e-inclusion programmes
Fields of interest
Communication for vulnerable audience. Digital communication. Health communication *Recent work*
Brecha digital y mayores (PROVULDIG-CM), reference H2015/HUM-3434, directed by Leopoldo Abad Alcalá/Ignacio Blanco (funded by the Dirección General de Universidades e Investigación de la Consejería de Educación, Juventud y Deporte (CAM)

Personas mayores, e-commerce y administracion electrónica: hacia la ruptura de la tercera brecha digital, reference CSO2015-66746-R, directed by Leopoldo Abad Alcalá (funded by the Convocatoria 2015—Proyectos I+D+I. Programa Estatal de Investigación, Desarrollo e Innovación (MINECO))

Auctoritas doméstica, capacitación digital y comunidad de aprendizaje en familias con menores escolarizados, reference CSO2013-42166-R, directed by Luis Nuñez Ladeveze y Teresa Torrecillas Lacave (funded by Convocatoria 2013—Proyectos I +D+I. Programa Estatal de Investigación, Desarrollo e Innovación (MINECO))

Comunidad escolar 2.0 La familia y la escuela ante los retos de la cultura digital. Diagnóstico y propuestas de actuación, reference FUSP-BS-PPC19/2014, directed by Tamara Vázquez Barrio (funded by Fundación CEU San Pablo)

Publications (selection)

Sanders K, Sánchez-Valle M, Viñaras M et al. (2015) Do we trust and are we empowered by "Dr. Google"? Older Spaniards' uses and views of digital healthcare communication. In: Public Relations Review 41(5):794–800. doi:10.1016/j.pubrev.2015.06.015

Llorente-Barroso C, Viñaras-Abad M, Sánchez-Valle M (2015) Mayores e Internet: La Red como fuente de oportunidades para un envejecimiento active. [Internet and the Elderly: Enhancing Active Ageing.] In: Comunicar 45:29–36. doi:10.3916/C45-2015-03

Frutos Torres B de, Pretel J, Sánchez-Valle M (2014) La interacción con las marcas en las redes sociales en jóvenes: hacia una presencia consentida y deseada, Adcomunica (7):69–89

Frutos Torres B de, Sánchez Valle M, Vázquez Barrio T (2014) Perfiles de adolescentes online y su comportamiento en el medio interactivo. Icono (14):374–397. doi:10.7195/ri14.v11i2.208

Further information and contact

www.uspceu@ceu.com

mvalle.fhum@ceu.es

Stefan Schmidt

M.Sc. health sciences and nursing, graduate of and doctoral student at the Martin-Luther-University, Halle-Wittenberg, research assistant at the department of social work and education of the University of Applied Sciences Neubrandenburg (D)

Book contribution

Caring TV—for older people with multimorbidity living alone—Positive feedback from users in Berlin and rural Mecklenburg-West Pomerania

Fields of interest

Technical assistance for frail and old people, care and case management

Recent work

Evaluation of Care Support Points of Mecklenburg-West Pomerania (2013–2015). (Ministry of Employment, Gender Equality and Social Affairs of

Mecklenburg-West Pomerania) SaLSA (2011–2014) Self-determined and Activated Life and Security in Old Age—a pilot study on the integration of Caring TV into the elderly support sector (Federal Ministry of Education and Research)
Personal Budget (2009–2010) (Excellency Program of Mecklenburg-West Pomerania)
Publications (selection)
Kollak I, Schmidt S (2016) Instrumente des Care und Case Management Prozesses. Springer Verlag, Berlin, Heidelberg
Jaccarini GA, Kollak I, Schmidt S (2015) Just in Case: Care and Case Management in Malta. MMDNA
Kollak I, Schmidt S (2015) Fallübungen Care und Case Management. Springer Verlag, Berlin, Heidelberg
Kampmeier AS, Kraehmer S, Schmidt S (eds) (2014) Das Persönliche Budget. Selbständige Lebensführung von Menschen mit Behinderungen. Kohlhammer Verlag, Stuttgart
Further information and contact
www.hs-nb.de
sschmidt@hs-nb.de

Helen Schneider

M.A., social worker, doctoral student, research assistant, Fachbereich Wirtschafts- und Sozialwissenschaften am RheinAhrCampus Remagen, University of Applied Sciences Koblenz (D)
Book contribution
Use and development of new technologies in public welfare services –A user-centred approach using step by step communication for problem solving
Fields of interest
Cultural sensitivity, quality management, psychomotoric and autonomy in nursing care
Recent grants and work
„Wenn Auswanderer alt werden"—Eine Studie über innovative Altenpflegeheimkonzepte am Beispiel eines deutsch-kanadischen Altenpflegeheims (2014) (Research centre University of Applied Sciences Koblenz)
E-Learning-Projekt am RheinAhrCampus im Fachbereich Wirtschafts- und Sozialwissenschaften (2014) (Department of development and quality management University of Applied Sciences Koblenz)
„Das Westend bewegt sich (aufeinander zu)"—Ein Projekt zur sozialen Integration durch Sport (2010–2013) (Hessian Ministry of Economy, Transport, Urban and Regional Development)
Publications (selection)
Schneider H (2015) Seniorengerechte Verpflegung sicherstellen. In: AOK-Verlag: PDCA—Qualitätsmanagement in der Pflege. AOK-Verlag, Remagen, no page references (loose-leaf collection)

Schneider H (2015) Innovationsbarrieren und Widerstände überwinden. In: Moos G, Peters A (eds) Innovationen in der Sozialwirtschaft. Nomos Verlag, Baden-Baden, pp 301–314
Schneider H (2013) Konstruktionen und Rekonstruktionen zum Bedarf von Altenpflegeheimbewohnern. In: Gahleitner SB, Hahn G, Glemser R (eds) Psychosoziale Interventionen. Klinische Sozialarbeit [Psychosocial interventions. Clinical social work], vol 6. Psychiatrie Verlag, Köln, pp 188–202
Further information and contact
schneider4@hs-koblenz.de

Laura Schönijahn

B.Sc., student assistant, Institut für Rehabilitationswissenschaften, Abt. Rehabilitationstechnik, Neue Medien, Humboldt University of Berlin (D)
Book contribution
Telemonitoring in home care—Creating the potential for a safer life at home
Fields of interest
Human-machine interaction, medical technology, neuroergonomics
Further information and contact
laura.schoenijahn@gmail.com

Barbara Steiner

Dr., head of the division old people's care, BruderhausDiakonie, Reutlingen, lecturer of social gerontology at Protestant University of Applied Sciences Ludwigsburg (D)
Book contribution
Use and development of new technologies in public welfare services- A user-centred approach using step by step communication for problem solving
Fields of interest
Housing and living arrangements in old age, civil engagement
Recent work
Assistent (2016–2018) Assisticrte Kommunikation in ambulanten Betreuungsformen, Innovationsprogramm Pflege (Ministry of Social Affairs and Integration Baden-Württemberg)
EmAsIn (2015–2018) Emotionssensitive Assistenzsysteme zur reaktiven psychologischen Interaktion mit Menschen (German Ministry of Education and Research)
PflegeCoDe (2015–2018). PflegeCoDe (2015–2018) Pflegecoaching für die optimale Unterstützung von Menschen mit Demenz (German Ministry of Education and Research) Living BaWüPrimer (2015) Smart Home and Living Baden-Württemberg (Ministry of Finance and Economy Baden-Württemberg
NiviL (2014–2017) nichtvisuelle Wirkung von Licht (German Ministry of Education and Research)
Publications (selection)

Steiner B (2015) „Unterstütztes Wohnen" für ältere Menschen mit Hilfe- und Pflegebedarf als neues Paradigma? Aspekte der Lebensqualität kleingemeinschaftlicher Wohnformen (dissertation), University of Heidelberg

Pfister V, Steiner B (2015) Durch Interdisziplinarität zur AAL-Lösung für Senioren? Ergebnisse einer Bedarfsanalyse für eine individualisierte Assistenzplattform. VDE Verlag, Berlin

Steiner B (2014) Technikunterstützung in der Praxis—Erfahrungen mit AAL-Lösungen. In: Technik hilft Wohnen. Wie wirken sich technische Hilfen im Alltag aus? KVJS (ed) Stuttgart 2014

Steiner B (2014) Mitarbeiter gezielt entwickeln. Der Qualifikationsrahmen für den Beschäftigungsbereich der Pflege und persönlichen Assistenz älterer Menschen in der Praxis. J Altenheim (2)

Wahl HW, Steiner B (2013) Innovative Wohnformen. In: Pantel J, Schröder J, Sieber C et al. (eds) Praxishandbuch der Altersmedizin. Geriatrie—Gerontopsychiatrie—Gerontologie. Kohlhammer Verlag, Stuttgart

Further information and contact
www.bruderhausdiakonie.de
barbara.steiner@bruderhausdiakonie.de

Mónica Vinarás Abad

Dr., adjunct professor in the department of audiovisual communication and publicity at the Universidad CEU San Pablo in Madrid (E)

Book contribution
Empowering the elderly and promoting active ageing through the Internet—The benefit of e-inclusion programmes

Fields of interest
Corporate and strategic communication, sponsorship public relations, crisis communication, corporate social responsibility, digital communication in museums

Recent work
Comunicación digital en las instituciones sanitarias para un envejecimiento activo, reference USPBSPPC03/2012, directed by Karen Sanders (funded by the University CEU San Pablo)

Brecha digital y mayores (PROVULDIG-CM), reference H2015/HUM-3434, directed by Leopoldo Abad Alcalá/Ignacio Blanco (funded by the Dirección General de Universidades e Investigación de la Consejería de Educación, Juventud y Deporte (CAM)

Personas mayores, e-commerce y administracion electrónica: hacia la ruptura de la tercera brecha digital, reference CSO2015-66746-R, directed by Leopoldo Abad Alcalá (funded by the Convocatoria 2015—Proyectos I+D+I. Programa Estatal de Investigación, Desarrollo e Innovación (MINECO))

Publications (selection)
Bartolomé A, Viñaras M, Alonso H (2016) El papel de la universidad en la RSC de las empresas. Option, Especial (4) vol 31, pp 686, 709

Sanders K, Sánchez-Valle M, Viñaras M et al. (2015) Do we trust and are we empowered by "Dr. Google"? Older Spaniards' uses and views of digital healthcare communication. In: Public Relations Review 41(5):794–800. doi:10.1016/j.pubrev. 2015.06.015

Viñaras M, Cabezuelo F, Herranza JM (2015) Corporate Philosophy and Values Brand New Paradigm as Axles Communicative. Prisma Social (14):379–410

Further information and contact

www.uspceu@ceu.com

monica.vinarasabad@ceu.es

Michael Wahl

Prof. Dr. Institut für Rehabilitationswissenschaften, Abt. Rehabilitationstechnik, Neue Medien, Humboldt University Berlin (D)

Book contribution

Telemonitoring in home care—Creating the potential for a safer life at home

Fields of interest

Tablet-PCs in der Rehabilitation, Erfassung von Blickbewegungen mittels Eyetrackingverfahren, Neue Technologien in der Rehabilitation, UK in der Sprachtherapie, Diagnostik und Neue Medien in der UK/

Publications (selection)

Jankowski N, Gerstmann J, Wahl M (2016) Nutzungsbereitschaft von Telemedizin in der Schlaganfallnachsorge—Sicht der Behandler. In: Schug S, Schmücker P, Semler S et al. (eds) E-Health-Rahmenbedingungen im europäischen Vergleich: Strategien, Gesetzgebung, Umsetzung. AKAVerlag, Berlin, pp 133–142

Wahl M, Wiedecke J (2015) Der Einsatz des iPads/Tablets im Unterricht bei Schülerinnen und Schülern mit sonderpädagogischem Förderbedarf: Eine Befragung. Zeitschrift für Heilpädagogik (4).191–205

Wahl M, Jankowski N, Klausner M et al. (2015) Neue Medien und Telemedizin in der Bewegungsrehabilitation—Chancen, Potentiale und Grenzen aus Sicht der beteiligten Personen. In: Gramann K, Zander TO, Wienrich C et al. (eds) Tagungsband zur 11. Berliner Werkstatt Mensch-Maschine-Systeme. Technische Universität, Berlin, pp 241–244

Speth F, Seifert U, Wahl M (2014) Robot-assisted arm training and Rhythmic Acoustic Stimulation: The role of different rhythmic stimuli on motivation and performance in motor-rehabilitation training after stroke. In: Song MK (ed) Proceedings of the ICMPC13-APSCOM5 2014 Joint Conference. Yonsei University, Seoul, pp 220–222

Marzinzik F, Wahl M, Klostermann F (2009) Ambulante videounterstützte Parkinsontherapie. Neuro Transmitter 20(9):16–18

Further information and contact

https://www.reha.hu-berlin.de/personal/mitarbeiter/1685900

michael.arnold-wahl@hu-berlin.de

Literature

Journals and Articles

Aas IHM (2001) A qualitative study of the organizational consequences of telemedicine. J Telemed Telecare 7(1):18–26

Alwin J, Persson J, Krevers B (2013) Perception and significance of an assistive technology intervention—the perspectives of relatives of persons with dementia. Disabil Rehabil 35 (18):1519–1526. doi:10.3109/09638288.2012.743603

Andrews J (2012) How acute care managers can support patients with dementia. J Nurs Manag 19 (2):18–20. doi:10.7748/nm2012.05.19.2.18.c9062

Andrews J (2012) A nurse manager's guide to support patients with dementia. Int J Older People Nurs 24(6):18–20

Astell AA (2006) Technology and personhood in dementia care. Quality in Ageing—Policy, practice and research 7(1):15–25. doi:10.1108/14717794200600004

Bantry E, Montgomery P (2014) Electronic tracking for people with dementia: An exploratory study of the ethical issues experienced by carers in making decisions about usage. Dementia 13:216–232. doi:10.1177/1471301212460445

Bartl-Pokorny KD, Pokorny F, Bölte S et al (2013) Eye-Tracking: anwendung in Grundlagenforschung und klinischer Praxis. Klin Neurophysiol 44(03):193–198 doi:10. 1055/s-0033-1343458

Bharucha AJ, London AI, Barnaid D et al (2006) Ethical considerations in the conduct of electronic surveillance research. J Law Med Ethics 34(3):611–619

Blanco Pérez A, Gutiérrez Couto U (2002) Legibilidad de las páginas web sobre salud dirigidas a pacientes y lectores de la población general. Revista española de salud pública 76(4):321–331

Borgestig M, Sandqvist J, Parsons R et al (2016) Eye gaze performance for children with severe physical impairments using gaze-based assistive technology: A longitudinal study. Assist Technol 28(2):93–102. doi:10.1080/10400435.2015.1092182

Breidenstein G, Hirschauer S (2002) Endlich fokussiert? Weder „Ethno" noch „Graphie". sozialer sinn. Zeitschrift für hermeneutische Sozialforschung 3(1):125–128

Cahill A, Begley E, Faulkner JP et al (2007) 'It gives me a sense of independence'—findings from Ireland on the use and usefulness of assistive technology for people with dementia. Technology and Disability 19:133–142

Cahill S, Macijauskiene J, Nygård AM et al (2007) Technology in dementia care. Technology and Disability 19:55–60

Callén B, Domènech M, López D et al (2009) Telecare research: (cosmo)politicizing methodology. ALTER 3(2):110–122. doi:10.1016/j.alter.2009.02.001

Corbetta D, Sirtori V, Castellini G et al (2015) Constraint-induced movement therapy for upper extremities in people with stroke. Cochrane Database Syst Rev 8(10):CD004433. doi:10.1002/14651858.CD004433.pub3

I. Kollak (ed.), *Safe at Home with Assistive Technology*,
DOI 10.1007/978-3-319-42890-1

Dybwik K, Tollali T, Nielsen EW et al (2011) "Fighting the system": families caring for ventilator-dependent children and adults with complex health care needs at home. BMC Health Serv Res. 11:156−164. doi:10.1186/1472-6963-11-156

Ehlert U (2011) Einen Augenblick, bitte! Unterstützte Kommunikation 4:14–18

Elsbernd A, Lehmeyer S, Schilling U (2012) Technik und Pflege—aktuelle Diskussionen und notwendige Entwicklungen, Pflegewissenschaft 9:453–458. doi:10.3936/1168

Fager S, Bardach L, Russell S et al (2012) Access to augmentative and alternative communication: new technologies and clinical decision-making. J Pediatr Rehabil Me 5(1):53–61. doi:10.3233/PRM-2012-0196

Flandorfer P (2012) Population ageing and socially assistive robots for elderly persons: the importance of sociodemographic factors for user acceptance. Int J Popul Res. doi:10.1155/2012/829835

Frutos Torres B de, Pretel J, Sánchez-Valle M (2014) La interacción con las marcas en las redes sociales en jóvenes: hacia una presencia consentida y deseada, Adcomunica (7):69–89

Frutos Torres B de, Sánchez Valle M, Vázquez Barrio T (2014) Perfiles de adolescentes online y su comportamiento en el medio interactivo. Icono (14):374–397

García-Guardia ML, Llorente-Barroso C (2016) An empirical study of the recreational-expressive and referential roles of the cell phone among young students and their potential applications to advertising. Prisma Social, Especial (1):261–278

Gastreich D (2013) Augenblick mal: Technik und künftige Anwendungsmöglichkeiten der Blicksteuerung. C't Magazin für Computer-Technik 25:168–175

Grans AL, Saborowski M, Kollak I (2016) Sprechen, Lachen, Rückwärtsfahren—was Kommunikationshilfen im Alltag erleben. Unterstützte Kommunikation (2):44–46

Grans AL, Wahl M (2013) Unterstützte Kommunikation—eine (neue) Profession? Ein Beitrag zur aktuellen Debatte um Professionalität und Professionalisierung in diesem pädagogischen Handlungsfeld. Zeitschrift für Heilpädagogik 64(11):480–485

Heaton J, Noyes J, Sloper P et al (2005) Families' experiences of caring for technology-dependent children: a temporal perspective. Health Soc Care Community 13(5):441–450. doi:10.1111/j.1365-2524.2005.00571.x

Hinkelmann J (2011) Anspruch auf das technisch Machbare. Unterstützte Kommunikation 1:33–35

Hischauer S (2001) Ethnografisches Schreiben und die Schweigsamkeit des Sozialen. Zu einer Methodologie der Beschreibung. Z Soziol 30(6):429–451

Holzhausen M, Martus P (2012) Validation of a new patient-generated questionnaire for quality of life in an urban sample of elder residents. Qual Life Res 22(1):131–135. doi:10.1007/s11136-012-0115-9

Horneber A (2011) Vom Segen einer Augensteuerung. Unterstützte Kommunikation 4:24–29

Huber JM, Eckstein C, Riedel A et al (2016) Bildungsübergänge durch Tutorien erfolgreich gestalten. Den Aufbau von Handlungskompetenz prinzipienorientiert und durch Reflexion in Peergruppen begleiten. PADUA Fachzeitschrift für Pflegepädagogik, Patienteneducation und -bildung 11(1):45–51

Hughes AM, Hallewell E, Kutlu M et al (2014) Combining electrical stimulation mediated by iterative learning control with movement practice using real objects and simulated tasks for post-stroke upper extremity rehabilitation. Neurosonology Cereb Hemodynamics 10(2):117–122

Hughes JC, Louw SJ (2002) Electronic tagging of people with dementia who wander. BMJ 325(7369):847–848

Johnson L, Burridge JH, Demain SH (2013) Internal and external focus of attention during gait re-education: an observational study of physical therapist practice in stroke rehabilitation. Phys Ther 93(7):957–966. doi:10.2522/ptj.20120300

Joronen K, Konu A, Rankin SH et al (2012) An evaluation of a drama program to enhance social relationships and anti-bullying at elementary school: a controlled study. Health Promotion International 27(1):5–14. doi:10.1093/heapro/dar012

Joronen K, Häkämies A, Åstedt-Kurki P (2011) Children's experiences of drama programme of social and emotional learning. Scandinavian Journal of Caring Sciences (25):671–678. doi:10.1111/j.1471-6712.2011.00877.x

Kääriäinen M, Lahtinen M (2006) Systemaattinen kirjallisuuskatsaus tutkimustiedon jäsentäjänä. Hoitotiede 18(1):37–45

Kampmeier AS, Kraehmer K (2014) Inklusion als Chance für Hochschulen und Hochschuldidaktik. Die Neue Hochschule. DNH. Heft 4—Jahrgang 2014:110–113

Kent-Walsh J, McNaughton D (2005) Communication partner instruction in AAC: Present practices and future directions. Augmentative and Alternative Communication 21:195–204

Knorz G (2015): Verständigung mit Hindernissen. Unterstützte Kommunikation 1:40–43

Knorz G, Saborowski M, Grans AL et al (2015) Entschleunigte Kommunikation oder beschleunigtes Missverstehen? alice—das Hochschulmagazin der ASH Berlin (29):57–59

Konu A, Joronen K, Lintonen T (2014) Seasonality in school well-being: the case of Finland. Child Indicators Research 8(2):265–277. doi:10.1007/s12187-014-9243-9

Lahtinen M, Rantala A, Heino-Tolonen T et al (2016) Lääkkeetöntä kivunlievitystä edistävät ja estävät tekijät lasten sairaalahoidon aikana. [Factors enhancing or hindering the use of non-pharmacological pain relief methods among hospitalized children.] Tutkiva Hoitotyö 14 (2):4–13

Lahtinen M, Rantala A, Heino-Tolonen T et al (2015) Lääkkeetön kivunlievitys ja sen kirjaaminen lasten sairaalahoidon aikana. [Non-pharmacological pain relief and its documentation among hospitalized children.] Hoitotiede 27(4):324–337

Lahtinen M, Joronen K (2014) Vanhempien kokemukset hengityslaitetta tarvitsevan lapsen hoitamisesta kotona—kirjallisuuskatsaus. [Parents' experiences of caring for ventilator-dependent child - a literature review.] Hoitotiede 26(2):89–100

Lai HH, Lin YC, Yeh CH et al (2006) User-oriented design for the optimal combination on product design. Int J Prod Econ 100(2):253–267. doi:10.1016/j.ijpe.2004.11.005

Landau R, Auslander GK, Shoval N et al (2010) Families and professional caregivers' views of using advanced technology to track people with dementia. Qual Health 20(3):409–419. doi:10.1177/1049732309359171

Landau R, Werner S, Auslander GK et al (2009) Attitudes of family and professional care-givers towards the use of GPS for tracking patients with dementia: an exploratory study. Br J Soc 39 (4):670–692. doi:10.1093/bjsw/bcp037

Lepistö S, Joronen K, Åstedt-Kurki P et al (2012) Subjective well-being in Finnish adolescents experiencing family violence. Journal of Family Nursing 18(2):200–233. doi:10.1177/1074840711435171

Liao LR, Lau RW, Pang MY (2012) Measuring environmental barriers faced by individuals living with stroke: development and validation of the chinese version of the Craig Hospital Inventory of Environmental Factors. J Rehabil Med 44(9):740–746. doi:10.2340/16501977-1014

Llorente-Barroso C, Viñarás-Abad M, Sánchez-Valle M (2015) Internet and the elderly: enhancing cctive ageing. Comunicar 45:29–36. doi:10.3916/C45-2015-03

Luebke N, Meinck M, von Renteln-Kruse W (2004) The Barthel Index in geriatrics. A context analysis fort he Hamburg Classification Manual. Z Gerontol Geriatr 37(4):316–326

Marzinzik F, Wahl M, Klostermann F (2009) Ambulante videounterstützte Parkinsontherapie. NeuroTransmitter 20(9):16–18

McCann D, Bull R, Winzenberg T (2015) Sleep deprivation in parents caring for children with complex needs at home: a mixed methods systematic review. J Fam Nurs 21(1):86–118. doi:10.1177/1074840714562026

McShane R, Gedling K, Keene J et al (1998) Getting lost in dementia. A longitudinal study of a behavioural symptom. Int Psychogeriatr 10(3):253–260

Meiland FJM, Batink BJJ, Overmars-Marx T et al (2014) Participation of end users in the design of assistive technology for people with mild to severe cognitive problems: The European Rosetta project. Int Psychogeriatr 26(5):769–779. doi:10.1017/S1041610214000088

Miller EL, Murray L, Richards L et al (2010) Comprehensive overview of nursing and interdisciplinary rehabilitation care of the stroke patient: a scientific statement from the American Heart Association. Stroke 41(10):2402–2448. doi:10.1161/STR.0b013e3181e7512b

Mokhtari M, Aloulou H, Tiberghien T et al (2012) New trends to support independence in persons with mild dementia—a mini-review. Gerontology 58(6):554–563. doi:10.1159/000337827

Newton L, Dickinson C, Gibson G et al (2016) Exploring the views of GPs, people with dementia and their carers on assistive technology: a qualitative study. BMJ Open 6:e011132. doi:10. 1136/bmjopen-2016-011132

Nicholas DB, Keilty K (2007) An evaluation of dyadic peer support for caregiving parents of children with chronic lung disease requiring technology assistance. Soc Work Health Care 44 (3):245–59. doi:10.1300/J010v44n03_08

Nudo R (2003) Adaptive plasticity in motor cortex: implications for rehabilitation after brain injury. J Rehabil Med 35(41 Suppl):7–10

Nuß C, Saborowski M (2014) Neue Hilfsmittel in der Unterstützten Kommunikation. Das Forschungsprojekt EyeTrack4all stellt sich vor. alice—das Hochschulmagazin der ASH Berlin (27):38

Nygård L, Starkhammar S (2007) The use of everyday technology by people with dementia living alone: mapping out the difficulties. Aging Ment Health 11(2):144–155

Orpwood R, Sixsmith A, Torrington J et al (2007) Designing technology to support quality of life of people with dementia. Technology and Disability 19:103–112

Parette P, Scherer M (2004) Assistive technology use and stigma. Education and Training in Developmental Disabilities 39(3):217–226

Peine A, Neven L (2011) Social-structural lag revisited. Gerontotechnology 10(3):129–139. doi:10.4017/gt.2011.10.3.002.00

Perry J, Beyer S, Holm S (2009) Assistive technology, telecare and people with intellectual disabilities: ethical considerations. J Med Ethics 35(2):81–86. doi:10.1136/jme.2008.024588

Räsänen T, Lintonen T, Joronen K et al (2015) Girls and boys gambling with health and well-being. Journal of School Health 85(4):214–222

Remers H (2010) Environments for ageing, assistive technology and self-determination: ethical perspectives. Inform Health Soc Care 35(3–4):200–210. doi:10.3109/17538157.2010.528649

Riedel A, Kimmerle B, Bonse-Rohmann M et al (2015) Spannungsfelder am Übergang von der beruflichen Bildung und Praxis an die Hochschule. In: Pädagogik der Gesundheitsberufe 2 (1):25–35

Riedel A, Huber JM, Linde AC (2013) Wiederkehrende ethische Dilemmata strukturiert reflektieren—Psychiatrische Pflegepraxis. CNE Schwerpunkt Beratung. PsychPflege 19:261–268

Robinson L, Hutchings D, Corner L et al (2007) Balancing rights and risks: conflicting perspectives in the management of wandering in dementia. Health, Risk & Society 9:389–406. doi:10.1080/13698570701612774

Saborowski M, Grans AL, Thedinga M et al (2016) „Dann würde ich versuchen, eine Augensteuerung zu beantragen."Befragung von Fachpersonen aus der Unterstützten Kommunikation zu Erfahrungen mit Augensteuerungen. Zeitschrift für Heilpädagogik (1):16–28

Saborowski M, Nuß C, Kollak I (2016) Kommunikationshilfsmittel mit Augensteuerung für nichtsprechende Personen mit schweren motorischen Einschränkungen. Abstracts der Vorträge des 6. Heilberufe Science Symposiums. HeilberufeSCIENCE (Suppl 7: 2–11):4–5. doi:10. 1007/s16024-016-0270-y

Saborowski M (2016) Zusammenfassung zu: „Dann würde ich versuchen, eine Augensteuerung zu beantragen.". Unterstützte Kommunikation (1):36–37

Saborowski M, Kollak I (2015) „How do you care for technology?"—Care professionals' experiences with assistive technology in care of the elderly. Technol Forecast Soc Change 93:133–140. doi:10.1016/j.techfore.2014.05.006

Saborowski M (2014) Vier Augen sehen mehr: Das Kooperationsprojekt EyeTrack4all präsentiert erste Ergebnisse. alice—das Hochschulmagazin der ASH Berlin (28):53

Steiner B (2014) Mitarbeiter gezielt entwickeln. Der Qualifikationsrahmen für den Beschäftigungsbereich der Pflege und persönlichen Assistenz älterer Menschen in der Praxis. J Altenheim (2)

Taub E, Uswatte G, Mark VW et al (2013) Method for enhancing real-world use of a more affected arm in chronic stroke: transfer package of constraint-induced movement therapy. Stroke 44 (5):1383–1388. doi:10.1161/STROKEAHA.111.000559

Tedesco Triccas L, Burridge JH, Hughes AM et al (2015) Multiple sessions of transcranial direct current stimulation and upper extremity rehabilitation in stroke: a review and meta-analysis. Clin Neurophysiol 127(1):946–955. doi:10.1016/j.clinph.2015.04.067

Tedesco Triccas L, Burridge JH, Hughes AM et al (2015) A double-blinded randomised controlled trial exploring the effect of anodal transcranial direct current stimulation and uni-lateral robot therapy for the impaired upper limb in sub-acute and chronic stroke. Neurorehabilitation: 1–28

Viñaras M, Cabezuelo F, Herranza JM (2015) Corporate philosophy and values brand new paradigm as axles communicative. Prisma Social (14):379–410

Wahl M, Wiedecke J (2015) Der Einsatz des iPads/Tablets im Unterricht bei Schülerinnen und Schülern mit sonderpädagogischem Förderbedarf: Eine Befragung. Zeitschrift für Heilpädagogik (4):191–205

Wang KW, Barnard A (2008) Caregivers' experiences at home with a ventilator-dependent child. Qual Health Res 18(4):501–508. doi:10.1177/1049732307306185

Wee SK, Hughes AM, JH Burridge JH et al (2015) Effect of trunk support on upper extremity function in people with chronic stroke and healthy controls. Phys Ther 95(8):1163–1171 doi:10.2522/ptj.20140487

Wee SK, Hughes AM, Burridge JH et al (2014) Trunk restraint to promote upper extremity recovery in stroke patients: a systematic review and meta-analysis. Neurorehabil Neural Repair 28(7):660–677. doi:10.1177/1545968314521011

Welsh S, Hassiotis A, O'Mahoney G et al (2003) Big brother is watching you—the ethical implications of electronic surveillance measures in the elderly with dementia and in adults with learning difficulties. Aging Ment Health 7(5):372–375

Winance M (2006) Trying out the wheelchair: the mutual shaping of people and devices through adjustment. Sci Technol Human Values 31(1).2–72. doi:10.1177/0162243905280023

Wu CY, Chen YA, Lin KC et al (2012) Constraint-induced therapy with trunk restraint for improving functional outcomes and trunk-arm control after stroke: a randomized controlled trial. Phys Ther 92(4):483–492. doi:10.2522/ptj.20110213

Books and Chapters

Akrich M (1992) The de-scription of technical objects. In: Bijker W, Law J (eds) Shaping technology—building society. Studies in sociotechnical change. MIT Press, Cambrigde, p 205–224

Akrich M, Latour B (1992) A summary of convenient vocabulary for the semiotics of human and nonhuman assemblies. In: Bijker W, Law J (eds) Shaping technology—building society. Studies in sociotechnical change. MIT Press, Cambrigde, p 259–264

Alzheimer's Society (2008) Dementia: out of the shadows. Alzheimer's Society, London

Andres P (1996) Die Bedeutung der Positionierung für eine erfolgreiche Unterstützte Kommunikation. In: von-Loeper-Literaturverlag, Isaac Gesellschaft für Unterstützte Kommunikation (eds) „Edi, mein Assistent" und andere Beiträge zur unterstützten Kommunikation. Reader der Kölner Fachtagungen. Verlag Selbstbestimmtes Leben, Düsseldorf, pp 290–299

Antener G (2001) Und jetzt? Das Partizipationsmodell in der Unterstützten Kommunikation. In: Boenisch J, Bünk C (eds) Forschung und Praxis der Unterstützten Kommunikation. von Loeper Literaturverlag, Karlsruhe, pp 257–267

Baldwin C (2008) Toward a person-centred ethic in dementia care: doing right or being good? In: Downs M, Bowers B (eds) Excellence in dementia care: research into practice. Open University Press, London

Baldwin C (2006) Reflections on ethics, dementia and technology. In: Woolham J (ed) Assistive technology in dementia care. Hawker publications, London

Baltes PB, Baltes MM (1990) Psychological perspectives on successful aging: the model of selective optimization with compensation. In: Baltes PB, Baltes MM (eds) Successful aging. Perspectives from behavioral sciences. Cambridge University Press, Cambridge, pp 1–34

Beauchamp TL, Childress JF (2001) Principles of biomedical ethics, 3rd edn. Oxford University Press, Oxford

Berker T, Hartmann M, Punie Y et al (2006) Domestication of media and technology. Open University Press, Maidenhead

Beukelman D, Mirenda P (1999) Augmentative and alternative communication. Management of severe communication disorders in children and adults, 2nd edn. P.H. Brookes Pub, Baltimore

Birngruber C (ed) (2009) Werkstatt Unterstützte Kommunikation. von Loeper Literaturverlag, Karlsruhe

Blackstone S, Hunt-Berg M (2006) Manual Soziale Netzwerke: Ein Instrument zur Erfassung der Kommunikation unterstützt kommunizierender Menschen und ihrer Kommunikationspartnerinnen und -partner. von Loeper Literaturverlag, Karlsruhe

Boenisch J (ed) (2005) Leben im Dialog: Unterstützte Kommunikation über die gesamte Lebensspanne. von Loeper Literaturverlag, Karlsruhe

Boenisch J (2003) Methoden der unterstützten Kommunikation. von Loeper Literaturverlag, Karlsruhe

Boenisch J (2001) Forschung und Praxis der unterstützten Kommunikation. von Loeper Literaturverlag, Karlsruhe

Boenisch J, Sachse S (2013) Diagnostik und Beratung in der Unterstützten Kommunikation: Theorie, Forschung und Praxis, 2nd edn. von Loeper Literaturverlag, Karlsruhe

Bollmeyer H (2011) UK inklusive: Teilhabe durch Unterstützte Kommunikation. von Loeper, Karlsruhe

Breidenstein G, Hirschauer S, Kalthoff H et al (2013) Ethnographie: Die Praxis der Feldforschung. UVK, Konstanz

Bruno J, Hansen F (2009) Diagnostiktest TASP: zur Abklärung des Symbol- und Sprachverhältnisses in der Unterstützten Kommunikation. Rehavista GmbH, Berlin

Castellot LR, Giuliano A, Mulvenna MD (2010) State of the art in electronic assistive technologies for people with dementia. In: Mulvenna MD, Nugent CD (eds) Supporting people with dementia using pervasive health technologies. Springer-Verlag, Heidelberg

Charmaz K (2010) Constructing Grounded Theory: A Pracitcal Guide Through Qualitative Analysis. Sage, Los Angelos et al

Debeljak M, Ocepek J, Zupan A (2012) Eye controlled human computer interaction for severely motor disabled children: two clinical case studies. In: Miesenberger K, Karshmer A, Penaz P et al (eds): Computers helping people with special needs: 13th International Conference, ICCHP 2012, Proceedings Part II. Springer, Berlin, New York, pp 153–156

De Larra RM (2004) Los Mayores en la Sociedad de la Información: situación actual y retos de futuro. In: De Larra RM (ed) Cuadernos Sociedad de la Información, vol 4. Fundacion AUNA, Madrid

Denecke J, Felix F, Pfister V et al (2016) Bedarfsanalysator zur Bestimmung eines Assistenzsystems (PATRONUS). VDE Verlag, Berlin

Deutscher Ethikrat (2011) Nutzen und Kosten im Gesundheitswesen—Zur normativen Funktion ihrer Bewertung. Stellungnahme. Deutscher Ethikrat, Berlin

Diefenbach S, Lenz E, Hassenzahl M (2013) An interaction vocabulary. Describing the how of interaction. In: ACM—Association for Computing Machinery (ed) CHI '13 Extended Abstracts on Human Factors in Computing Systems. ACM, New York, pp 607–612

Diekmann N, Sande KI, Steinhaus I (2007) Partnerbasierte Kommunikationsstrategien für Menschen mit schweren Beeinträchtigungen: ein Konzept von Linda Burkhart und Gayle Porter. In: Sachse S, Birngruber C, Arends S (eds) Lernen und Lehren in der Unterstützten Kommunikation. von Loeper Literaturverlag, Karlsruhe, pp 38–47

Dienel HL, Peine A, Cameron H (2004) New participative tools in product development for seniors. In: Burdick DC, Kwon S (eds) Gerotechnology. Research and practice in technology and aging. Springer Publishing, New York, pp 224–238

Elsbernd A (2013) Konzepte für die Pflegepraxis: Theoretische Einführung in die Konzeptentwicklung pflegerischer Arbeit. In: Einführung von ethischen Fallbesprechungen. Ein Konzept für die Pflegepraxis. Ethisch begründetes Handeln praktizieren, vol 3. Jacobs Verlag, Lage, pp 13–38

Erdmann B, Schweigert H (2012) Zahlungsbereitschaft für AAL-Produkte. Tagungungsband Technik für ein selbstbestimmtes Leben—5. Deutscher AAL-Kongress 24.01.2012–25.01.2012 in Berlin

Ericsson KA, Krampe RT, Tesch-Römer C (1993) The role of deliberate practice in the acquisition of expert performance, vol. 100. American Psychological Association, pp 363–406

Fitzpatrick G, Balaam M, Axelrod L et al (2010) Designing for rehabilitation at home. In: Gillian RH, Desney ST, Gillian H et al. (eds) Proceedings workshop on interactive systems in healthcare (WISH10), Atlanta April 11 2010, pp 49–52

Flick U, Kardorff von E, Steinke I (2013) Qualitative Forschung: Ein Handbuch, 10th edn. Rowohlt-Taschenbuch-Verlag, Reinbek bei Hamburg

Giel B (2014) Interdisziplinäre Zusammenkünfte: Grundlage einer teilhabeorientierten UK. In: von-Loeper-Literaturverlag, Isaac—Gesellschaft für Unterstützte Kommunikation (eds) Handbuch der Unterstützten Kommunikation. von Loeper Literaturverlag, Karlsruhe, pp 01.056.001–01.061.001

Gläser J, Laudel G (2009) Experteninterviews und qualitative Inhaltsanalyse als Instrumente rekonstruierender Untersuchungen, 3rd edn. VS Verlag für Sozialwissenschaften, Wiesbaden

Haddon L (2006) Empirical studies using the domestication framework. In: Berker T (ed) Domestication of media and technology. Open University Press, Maidenhead, pp 103–122

Hagen I, Cahill S, Begley E et al (2005) Assessment of usefulness of assistive technologies for people with dementia. In: Pruski A, Knops H (eds) Assistive Technology: From Virtuality to Reality. Assistive Technology Research Series, vol 16. IOS Press, Amsterdam et al., pp 348–352

Hallbauer A, Hallbauer T, Hüning-Meier M (eds) (2013) UK kreativ!: Wege in der unterstützten Kommunikation. von Loeper Literaturverlag, Karlsruhe

Häußler A (2005) Der TEACCH-Ansatz zur Förderung von Menschen mit Autismus. Einführung in Theorie und Praxis. Borgmann Media, Dortmund

Hedderich I (2006) Unterstützte Kommunikation in der Frühförderung: Grundlagen—Diagnostik—Beispiele. Klinkhardt, Bad Heilbrunn

Heim M, Jonker V, Veen M (2012) COCP: Ein Interventionsprogramm für nicht sprechende Personen und ihre Kommunikationspartner. In: von-Loeper-Literaturverlag, Isaac—Gesellschaft für Unterstützte Kommunikation (eds) Handbuch der Unterstützten Kommunikation, 5th edn. von Loeper Literaturverlag, Karlsruhe, pp 01.026.007–01.026.015

Hopf C (2005) Qualitative Interviews: ein Überblick. In: Kardorff von E, Steinke I, Flick U (eds) Qualitative Forschung. Ein Handbuch, 4th edn. Rowohlt, Reinbek bei Hamburg, pp 349–360

Hörmeyer I (2012) The importance of gaze in the constitution of units in Augmentative and Alternative Communication (AAC) In: Bergmann P, Brenning J, Pfeiffer M et al (eds) Prosody and embodiment in interactional grammar. de Gruyter, Berlin, pp 237–264

Jaccarini GA, Kollak I, Schmdit S (2015) Just in case: care and case management in Malta. MMDNA

Jankowski N, Gerstmann J, Wahl M (2016) Nutzungsbereitschaft von Telemedizin in der Schlaganfallnachsorge—Sicht der Behandler. In: Schug S, Schmücker P, Semler S et al (eds) E-Health-Rahmenbedingungen im europäischen Vergleich: Strategien, Gesetzgebung, Umsetzung. AKA Verlag, Berlin, pp 133–142

Jankowski N, Gerstmann J, Wahl M (2015) Unterstützungsansätze in der Versorgung mittels mobiler Geräte und Telemedizin. In: Weisbecker A, Burmester M, Schmidt A (eds) Mensch und Computer 2015—Workshopband. De Gruyter, Berlin, Boston, pp 117–124

Jankowski N et al (2014) Technologies in Practice. In: Bister M, Niewöhner J (eds) Alltag in der Psychiatrie im Wandel. Ethnographische Perspektiven auf Wissen, Technologie und Autonomie. Panama Verlag, Berlin

Johansson K (2007) Kirjallisuuskatsaukset—Huomio systemaattisen kirjallisuuskatsaukseen. In: Johansson K, Axelin A, Stolt M et al (eds) Systemaattinen kirjallisuuskatsaus ja sen tekeminen. Hoitotieteen laitoksen julkaisuja. Tutkimuksia ja raportteja A:51/2007. Turun yliopisto, Turku, pp 3–9

Kaminski A (2010) Technik als Erwartung. Transcript, Bielefeld

Kampmeier AS, Kraehmer S, Schmidt S (eds) (2014) Das Persönliche Budget. Selbständige Lebensführung von Menschen mit Behinderungen. Kohlhammer Verlag, Stuttgart

Kampmeier AS (2010) Transition zwischen den Paradigmen—Stolperstein—Persönliches Budget. In: Schildmann U (ed) Umgang mit Verschiedenheit in der Lebensspanne. Klinkhardt Verlag, Bad Heilbrunn

Kampmeier AS (2010) Realisierung des Persönlichen Budgets in der Hilfe für Menschen mit Behinderungen. In: Michel-Schwartze B (ed) "Modernisierungen" methodischen Handelns in der Sozialen Arbeit. VS Verlag, Wiesbaden

Kamps N (2014) UK-Hilfsmittelversorgung als Aufgabe der Gesetzlichen Krankenversicherung (GKV). In: von Loeper-Literaturverlag, Isaac—Gesellschaft für Unterstützte Kommunikation (eds) Handbuch der Unterstützten Kommunikation. von Loeper Literaturverlag, Karlsruhe, pp 16.003.001–16.015.001

Kauschke C (2000) Der Erwerb des frühkindlichen Lexikons. Eine empirische Studie zur Entwicklung des Wortschatzes im Deutschen. Gunter Narr Verlag, Tübingen

Kawai T, Takahashi H, Kamimura A et al (2011) Eye size recognition to control AAC devices for persons with Amyotrophic Lateral Sclerosis (ALS). In: Gelderblom GJ, Soede M, Adriaens L et al (eds) Assistive technology research series—everyday technology for independence and care. IOS Press, Amsterdam et al., pp 409–415

Kim HS (2015) The essence of nursing practice. Springer Publishing, New York

Kim HS, Schwartz-Barcott D, Holter IM (2012) Cross-cultural use and validity of pain scales and questionnaires—norwegian case study. In: Incayawar M, Todd KH (eds) Pain, culture, brain and analgesia—understanding and managing pain in diverse populations. Oxford University Press, New York

Kim HS (2012) Action science. In: Fitzpatrick J (ed) Encyclopedia of nursing research. Springer Publications, New York

Kim HS, Clabo LML, Burbank P et al (2010) Application of critical reflective inquiry in nursing education. In: Lyons N (ed) Handbook of reflection and reflective inquiry. Springer Science +Business Media, New York, pp 159–172

Kim HS (2010) The nature of theoretical thinking in nursing, 3rd ed. Springer Publishing, New York

Kimmerle B, Huber JM, Riedel A et al (2015) Pflegeberuflich Qualifizierte: Betrachtung einer Studierendengruppe beim Übergang in die Hochschule In: Freitag WK, Buhr R, Danzeglocke EM et al (eds) Übergänge gestalten. Durchlässigkeit zwischen beruflicher und hochschulischer Bildung erhöhen. Waxmann Verlag, Bonn, pp 151–172

Kitzinger A, Kristen U, Leber I (2010) Jetzt sag ich's Dir auf meine Weise: Erste Schritte in Unterstützter Kommunikation mit Kindern. 5th edn. von Loeper Literaturverlag, Karlsruhe

Kollak I (ed) (2016) Menschen mit Demenz durch Kunst und Kreativität aktivieren. Springer Verlag, Berlin, Heidelberg

Kollak I, Schmidt S (2016) Instrumente des Care und Case Management Prozesses. Springer Verlag, Berlin, Heidelberg

Kollak I, Schmidt S (2015) Fallübungen care und case management. Springer Verlag, Berlin, Heidelberg

Kollak I (2014) Time Out—Übungen zur Selbstsorge und Entspannung für Gesundheitsberufe. Springer Verlag, Berlin, Heidelberg

Kontopodis M, Niewöhner J (eds) (2011) Das Selbst als Netzwerk: zum Einsatz von Körpern und Dingen im Alltag. Transcript, Bielefeld

Kristen U (2012) Das Kommunikationsprofil: Ein Beratungs- und Diagnosebogen. In: von Loeper-Literaturverlag, Isaac—Gesellschaft für Unterstützte Kommunikation (eds) Handbuch der Unterstützten Kommunikation, 5th edn. von Loeper Literaturverlag, Karlsruhe, pp 12.017.001–12.038.001

Kristen U (2012) Diagnosebogen zur Abklärung kommunikativer Fähigkeiten In: von Loeper-Literaturverlag, Isaac—Gesellschaft für Unterstützte Kommunikation (eds) Handbuch der Unterstützten Kommunikation, 5th edn. von Loeper Literaturverlag, Karlsruhe, pp 14.023.001–14.030.001

Kristen U (2005) Praxis unterstützte Kommunikation: Eine Einführung, 5th edn. Verlag Selbstbestimmtes Leben, Düssedorf

Kruse A, Schmitt E, Holfelder JD et al (2011) Kreativität im Alter—Ergebnisse der Auswertung von Bewerbungen zum Otto-Mühlschlegel-Preis. In: Kruse A (ed) Kreativität im Alter. Universitätsverlag Winter, Heidelberg, pp 195–234

Kühn G, Schneider J (2009) Zwei Wege zur Kommunikation—Ein Praxisleitfaden zu TEACCH und PECS. Verlag hörgeschädigter Kinder gGmbH, Hamburg

Lage D (2006) Unterstützte Kommunikation und Lebenswelt. Eine kommunikationstheoretische Grundlegung für eine behindertenpädagogische Konzeption. Klinkhardt, Bad Heilbrunn

Leber I (2014) Kommunikation einschätzen und unterstützen: Poster und Begleitheft zu den Fördermöglichkeiten in der Unterstützten Kommunikation, 5th edn. von Loeper Literaturverlag, Karlsruhe

Lemler K, Gemmel S (2005) Kathrin spricht mit den Augen. Wie ein behindertes Kind lebt. edition zweihorn, Neureichenau

Majaranta P, Aoki H, Donegan M et al (eds) (2012) Gaze interaction and applications of eye tracking: advances in assistive technologies. IGI Global, Hershey, PA

Martin S, Bengtsson JE, Dröes RM (2010) Assistive technologies and issues relating to privacy, ethics and security. In: Mulvenna MD, Nugent CD (eds) Supporting people with dementia using pervasive health technologies. Springer-Verlag, Heidelberg, pp 63–76

Megges H, Jankowski N, Peters O (2013) Caregiver needs analysis for product development of an assistive technology system in dementia care. In: 23th Alzheimer Europe Conference, St. Julian's, Malta

Mialet H (2012) Hawking incorporated: Stephen Hawking and the anthropology of the knowing subject. The University of Chicago Press, Chicago

Mol A, Moser I, Pols J (2010) Care: putting practice into theory. In: Mol A, Moser I, Pols J (eds) Care in practice. On tinkering in clinics, homes and farms. transcript, Bielefeld, pp 7–25

Moser I, Law J (2003) Making voices'—new media technologies, disabilities, and articulation. In: Liestol G, Morrison A, Rasmussen T (eds) Digital media revisited: theoretical and conceptual innovation in digital domains. MIT Press, Cambridge, pp 491–520

Niedecken I, Hackstein J (2012) Recht auf Kommunikation: Ein Recht auf Unterstützung der Kommunikation. In von Loeper-Literaturverlag, Isaac—Gesellschaft für Unterstützte Kommunikation (eds) Handbuch der Unterstützten Kommunikation, 5th edn. von Loper Literaturverlag, Karlsruhe, pp 16.017.001–16.023.001

Oswald F (2002) Wohnbedingungen und Wohnbedürfnisse im Alter. In: Schlag B, Megel K (eds) Mobilität und gesellschaftliche Partizipation im Alter. Kohlhammer, Stuttgart, pp 97–115

Otto K, Wimmer B (2010) Unterstützte Kommunikation: Ein Ratgeber für Eltern, Angehörige sowie Therapeuten und Pädagogen, 3rd edn. Schulz-Kirchner, Idstein

Oudshoorn N (2011) Telecare technologies and the transformation of healthcare. Palgrave Macmillan, Houndmills et al

Oudshoorn N, Pinch T (2008) User-technology relationships: some recent developments. In: Hackett EJ (ed) The handbook of science and technology studies. MIT Press, Cambridge, pp 541–565

Pfeil S (2013) "Schaun mer mal.": Wege hin zum Augensteuerungskaiser! In: Hallbauer A, Hallbauer T, Hüning-Meier M (eds) UK kreativ! Wege in der unterstützten Kommunikation. von Loeper Literaturverlag, Karlsruhe, pp 390–395

Pfister V, Steiner B (2015) Durch Interdisziplinarität zur AAL-Lösung für Senioren? Ergebnisse einer Bedarfsanalyse für eine individualisierte Assistenzplattform. VDE Verlag, Berlin

Picot A, Braun G (eds) (2010) Telemonitoring in Gesundheits- und Sozialsystemen: Eine eHealth-Lösung mit Zukunft. Springer-Verlag, Berlin, Heidelberg

Pivit C, Hüning-Meier M (2012) Wie lernt ein Kind unterstützt zu kommunizieren? Allgemeine Prinzipien der Förderung und Prinzipien des Modelings. In: von LoeperLiteraturverlag, Isaac—Gesellschaft für Unterstützte Kommunikation (eds) Handbuch der Unterstützten Kommunikation, 5th edn. von Loeper Literaturverlag, Karlsruhe, pp 01.032.001–01.037.008

Pointner C, Malzer R (2011) Bei dir klickt's wohl? Lernsoftware in der Unterstützten Kommunikation. In: Bollmeyer H, Engel K, Hallbauer A et al (eds) UK inklusive—Teilhabe durch Unterstützte Kommunikation. von Loeper Literaturverlag, Karlsruhe, pp 118–129

Pudas-Tähkä SM, Axelin A (2007) Systemaattisen kirjallisuuskatsauksen aiheen rajaus, hakutermit ja abstraktien arviointi. In: Johansson K, Axelin A, Stolt M et al (eds) Systemaattinen kirjallisuuskatsaus ja sen tekeminen. Hoitotieteen laitoksen julkaisuja. Tutkimuksia ja raportteja A:51/2007. Turun yliopisto, Turku, pp 46−57

Renner G (2004) Theorie der unterstützten Kommunikation. Eine Grundlegung. Wiss.-Verl. Spiess, Berlin

Röser J, Peil C (2010) Teilhabe durch Domestizierung: Potenziale des häuslichen Alltags für den Zugang zum Internet. DGPuK-Fachgruppe Soziologie der Medienkommunikation, Frankfurt

Rothmayr A (2008) Pädagogik und unterstützte Kommunikation: Eine Herausforderung für die Aus- und Weiterbildung. von Loeper Literaturverlag, Karlsruhe

Saborowski M, Grans AL, Kollak I (2015) Wenn Blicke die Kommunikation steuern—Beobachtung einer Augensteuerung im Alltag. In: Antener G, Blechschmidt A, Ling K (eds) UK wird erwachsen. Initiativen in der Unterstützten Kommunikation. von Loeper Literaturverlag, Karlsruhe, pp 370–383

Saborowski M, Nuß C, Kollak I (2015) Taking a closer look at user-technology relationships: a network model. In: Gross M, von Klinski S (eds) Research Day 2015 „Stadt der Zukunft", Tagungsband 21.04.2015 der Beuth Hochschule Berlin. Mensch und Buch Verlag, Berlin, pp 125–129

Sachse S, Boenisch J (2012) Kern- und Randvokabular in der Unterstützten Kommunikation: Grundlagen und Anwendung. In: von Loeper-Literaturverlag, Isaac—Gesellschaft für Unterstützte Kommunikation (eds) Handbuch der Unterstützten Kommunikation, 5th edn. von Loeper Literaturverlag, Karlsruhe, pp 01.026.030–01.026.040

Sachse S, Willke M (2011) Fokuswörter in der Unterstützten Kommunikation. In: Bollmeyer H, Engel K, Hallbauer A et al (eds) UK inklusive: Teilhabe durch Unterstützte Kommunikation. von Loeper Literaturverlag, Karlsruhe, pp 375–394

Sachse S (2010) Interventionsplanung in der unterstützten Kommunikation: Aufgaben im Kontext der Beratung. von Loeper Literaturverlag, Karlsruhe

Sachse S (ed) (2007) Lernen und Lehren in der unterstützten Kommunikation. von Loeper Literatur Verlag, Karlsruhe

Schneider H (2015) Seniorengerechte Verpflegung sicherstellen. In: AOK-Verlag: PDCA—Qualitätsmanagement in der Pflege. AOK-Verlag, Remagen

Schneider H (2015) Innovationsbarrieren und Widerstände überwinden. In: Moos G, Peters A (eds) Innovationen in der Sozialwirtschaft. Nomos Verlag, Baden-Baden, pp 301–314

Schneider H (2013) Konstruktionen und Rekonstruktionen zum Bedarf von Altenpflegeheimbewohnern. In: Gahleitner SB, Hahn G, Glemser R (eds) Psychosoziale Interventionen. Klinische Sozialarbeit [Psychosocial interventions. Clinical social work], vol 6. Psychiatrie Verlag, Köln, pp 188–202

Schuh A (2013) Kommunikation und Computerzugang im 21. Jahrhundert: Wie kommunizieren wir heute? - am Beispiel der Augensteuerung Tobii PC eye. In: Hallbauer A, Hallbauer T, Hüning-Meier M (eds) UK kreativ! Wege in der unterstützten Kommunikation. von Loeper Literaturverlag, Karlsruhe, pp 396–408

Seiler-Kesselheim A (2008) Beratungsangebote in der Unterstützten Kommunikation: Praxis, Forschung, Weiterentwicklung. von Loeper Literaturverlag, Karlsruhe, Baden

Spiekermann A (2011) Altes Wissen in neuen Rechnern: Unterstützte Inklusion mit der Lese- und Schreibsoftware Multitext. In: Bollmeyer H, Engel K, Hallbauer A et al (eds) UK inklusive—Teilhabe durch Unterstützte Kommunikation. von Loeper Literaturverlag, Karlsruhe, pp 130–140

Staatsinstitut für Schulqualität und Bildungsforschung München (ISB) (ed) (2009) Unterstützte Kommunikation (UK) in Unterricht und Schule. Bayrisches Staatsministerium für Unterricht und Kultus München. Verlag Alfred Hintermaier, München

Staiger-Sälzer P, Veit S (2012) Blicktafeln: Wirkungsvolle nichttechnische Kommunikationshilfen in der Unterstützten Kommunikation. In: von-Loeper-Literaturverlag, Isaac—Gesellschaft für Unterstützte Kommunikation (eds) Handbuch der Unterstützten Kommunikation, 5th edn. von Loeper Literaturverlag, Karlsruhe, pp 05.013.001–05.016.001

Tetzchner S, Martinsen H (2000) Introduction to augmentative and alternative communication, 2nd edn. Whurr, London

Thygesen H, Moser I (2010) Technology and good dementia care: an argument for an ethics-in-practice approach. In: Schillmeier M, Domènech M (eds) New technologies and emerging spaces of care. Farnham, Ashgate, pp 129–147

Vickers S (2011) Eye-gaze interaction techniques for use in online games and environments for users with severe physical disabilites (PhD thesis). De Montfort University

Wachsmuth S (2006) Kommunikative Begegnungen. Aufbau und Erhalt sozialer Nähe durch Dialoge mit Unterstützter Kommunikation. Bentheim, Würzburg

Wahl M, Jankowski N, Klausner M et al (2015) Neue Medien und Telemedizin in der Bewegungsrehabilitation—Chancen, Potentiale und Grenzen aus Sicht der beteiligten Personen. In: Gramann K, Zander TO, Wienrich C et al (eds) Tagungsband zur 11. Berliner Werkstatt Mensch-Maschine-Systeme. Technische Universität, Berlin, pp 241–244

Wahl HW, Steiner B (2013) Innovative Wohnformen. In: Pantel J, Schröder J, Sieber C et al (eds) Praxishandbuch der Altersmedizin. Geriatrie—Gerontopsychiatrie—Gerontologie. Kohlhammer Verlag, Stuttgart

Weid-Goldschmidt B (2013) Zielgruppen unterstützter Kommunikation: Fähigkeiten einschätzen —Unterstützung gestalten. von Loeper Literaturverlag, Karlsruhe

Wiegerling K, Heesen J, Siemoneit O et al (2008) Ubiquitärer Computer—Singulärer Mensch. In: Klumpp D, Kubicek H, Rossnagel A et al (eds) Informationelles Vertrauen für die Informationsgesellschaft, Springer, Berlin, Heidelberg, pp 71–84

Witzner Hansen D, Majaranta P (2012) Basics of camera-based gaze tracking. In: Räihä KJ, Hyrskykari A, Hansen J et al (eds) Gaze interaction and applications of eye tracking. Advances in assistive technologies. IGI Global, Hershey, pp 21–26

World Health Organization (WHO) (ed) (2007) International classification of functioning, disability and health (ICF). WHO Press, Geneva

World Health Organization (2002) Reducing stigma and discrimination against older people with mental disorders; a technical consensus statement. World Health Organization Geneva

Zippel-Schultz B, Budych K, Schöne A et al (2013) With telemedical networks towards a better Case and Care Management for patients with chronical heart failure and rhythmic disorders. In: Haas P, Semmler SC, Schug SH et al (eds) Nutzung, Nutzer, Nutzen von Telematik in der Gesundheitsversorgung – eine Standortbestimmung, Tagungsband der TELEMED 2013. TMF, Berlin

Online Publications

Alzheimer's Association (2016) Alzheimer's disease facts and figures. Available via DIALOG. https://www.alz.org/documents_custom/2016-facts-and-figures.pdf. Accessed 06 Aug 2016

Alzheimer's Association (2016) Stages of Alzheimer's. Available via DIALOG. https://www.alz.org/alzheimers_disease_stages_of_alzheimers.asp?type=carecenter_footer. Accessed 06 Aug 2016

Alzheimer's Association (2015) Electronic tracking. Available via DIALOG. http://www.alz.org/documents_custom/statements/Electronic_Tracking.pdf. Accessed 06 Aug 2016

Alzheimer's Society (2008) Assistive technology. Available via DIALOG. www.alzheimers.org.uk. Accessed 06 Aug 2016

Bjørneby S, Topo P, Holthe T (1999) Technology, ethics and dementia: a guidebook on how to apply technology in dementia care. Norwegian Centre for Dementia Research, Oslo. Available via DIALOG. https://www.uni-bamberg.de/fileadmin/uni/fakultaeten/sowi_professuren/urbanistik/ted.pdf. Accessed 06 Aug 2016

Borgestig M (2016) The impact of gaze-based assistive technology on daily activities in children with severe physical impairments. Linköping University Medical Dissertations No. 1490. Available via DIALOG. http://www.akademiska.se/Global/KB/Folke%20Bernadotte%20regionhabilitering/Dokument/Spikblad.pdf. Accessed 31 Jul 2016

Bundesministerium für Arbeit und Soziales (2013) Das (trägerübergreifende) Persönliche Budget. Available via DIALOG. http://www.bmas.de/SharedDocs/Downloads/DE/PDF-Publikationen/a722-pers-budget-normalesprache.pdf?__blob=publicationFile. Accessed 31 Jul 2016

Bundesministerium für Arbeit und Soziales (2013) Das trägerübergreifende Persönliche Budget: Version in leichter Sprache. Available via DIALOG. http://www.bmas.de/SharedDocs/Downloads/DE/PDF-Publikationen/a722-pers-budget-einfachesprache.pdf?__blob=publicationFile. Accessed 31 Jul 2016

COGAIN: Project „Communication by gaze interaction (COGAIN)", WIKI with slides and texts. Available via DIALOG. http://wiki.cogain.org/index.php/Eye_Control_Hints_and_Tips. Accessed 18 Aug 2016

ENABLE (2004) Enabling technology for people with dementia: Report of the assessment study in England. Available via DIALOG. http://www.enableproject.org/download/Enable%20-%20National%20Report%20-%20UK.pdf. Accessed 08 Aug 2016

Goytia-Prat A, Lázaro-Fernández Y (eds) (2007) La experiencia de ocio y su relación con el envejecimiento activo. Instituto de Estudios de Ocio, Universidad de Deusto, Bilbao. Available via DIALOG. http://www.imserso.es/InterPresent1/groups/imserso/documents/binario/idi120_06udeusto.pd. Accessed 22 Aug 2016

Hoffmann L, Wülfing J-O (2010) Usability of electronic communication aids in the light of daily use. Presentation at the 14th Biennial Conference of the International Society for Augmentative and Alternative Communication, Barcelona, Spain, 24th - 29th July, 2010. Available via DIALOG. http://nbn-resolving.org/urn:nbn:de:0011-n-1457507. Accessed 31 Jul 2016

IMSERSO (2013) Envejecimiento Activo. Libro Blanco. Available via DIALOG http://www.
    imserso.es/InterPresent2/groups/imserso/documents/binario/8088_8089libroblancoenv.pd.
    Accessed 22 Aug 2016
International Organization for Standardization: ISO 9999 (2011) Assistive products for persons
    with disability—classification and terminology. Available via DIALOG. http://www.iso.org/
    iso/home/store/catalogue_ics/catalogue_detail_ics.htm?csnumber=50982. Accessed 14 May
    2013
Kollak I, Nuß C, Saborowski M (2016) Handreichung Augensteuerung: Hilfestellung für
    Vorüberlegungen, Planung und Einsatz einer Augensteuerung in der Unterstützten
    Kommunikation. Available via DIALOG. https://opus4.kobv.de/opus4-ash/frontdoor/index/
    index/docId/133. Accessed 23 Aug 2016
Living made easy (2010) Ethical issues with assistive technology. Available via DIALOG. http://
    www.livingmadeeasy.org.uk/scenario.php?csid=43. Accessed 08 Aug 2016
Luhmann N (2000) Familiarity, confidence, trust: problems and alternatives. In: Gambetta D
    (eds) Trust: making and breaking cooperative relations, University of Oxford, 94 107.
    Available via DIALOG. http://www.sociology.ox.ac.uk/papers/luhmann94-107.pdf. Accessed
    14 May 2013
Meagher C, Burridge J, Ewings S et al (2015) Feasibility randomised control trial of LifeCIT, a
    web-based support programme for Constraint Induced Therapy (CIT) following stroke
    compared with usual care. Physical Therapy: 1–24. Available via DIALOG. http://eprints.
    soton.ac.uk/384768/. Accessed 23 Aug 2016
Mental Welfare Commission for Scotland (2006) Rights, risks and limits to freedom. Mental
    Welfare Commission for Scotland. Available via DIALOG. http://hub.careinspectorate.com/
    media/110457/mwc-rights_risks_2013_edition.pdf. Accessed 08 Aug 2016
Mort M, Roberts C, Pols J et al (2013) Ethical implications of home telecare for older people: a
    framework derived from a multisited participative study. Available via DIALOG. http://dx.doi.
    org/10.1111/hex.12109. Accessed 10 Dez 2013
Pressmann H, Pietrzyk A (2012) Free and inexpensive apps for people who need augmentative
    communication supports. Available via DIALOG. http://www.centralcoastchildrensfoundation.
    org/draft/wp-content/uploads/2012/03/FreeandInexpensiveAACAppsFinal.pdf. Accessed 23
    Aug 2016
Saborowski M, Nuß C, Kollak I (2014) How to conceptualise AAC user technology relationships:
    a study on eye control. Postervortrag (Nr. 429) auf der ISAAC-Konferenz in Lissabon,
    Portugal, 22.07.2014. Available via DIALOG. https://www.isaac-online.org/wordpress/wp-
    content/uploads/EyeTrack-for-ISAAC_ASH-Berlin.pdf. Accessed 23 Aug 2016
Schiemann D, Moers M (2004) Werkstattbericht über ein Forschungsprojekt zur
    Weiterentwicklung der Methode „Stationsgebundene Qualitätsentwicklung in der Pflege".
    Available via DIALOG. https://www.dnqp.de/fileadmin/HSOS/Homepages/DNQP/Dateien/
    Weitere/WerkstattberichtSQE.pdf. Accessed 31 Jul 2016
Wey, S (n.d.). The ethical use of assistive technology. Available via DIALOG. https://www.
    atdementia.org.uk/content_files/files/The_ethical_use_of_assistive_technology.pdf. Accessed
    08 Aug 2016

Homepages

ALS-Selbsthilfe: Informationen für Betroffene, Angehörige (auch Kinder) und Fachpersonen zu
    ALS: http://www.als-selbsthilfe.de
Alzheimer`s Association:
www.alz.org
Alzheimer`s Community:
www.alzheimers.net
Alzheimer`s Foundation of America:

www.alzfdn.org

Andres P (2009): „Lernen mit Lennart" ein Vortrag über den Einsatz einer Blicktafel: https://www.youtube.com/watch?v=dlblMVK6sf0

AT dementia

www.atdementia.org.uk

CLUKS—Forum für Computergestütztes Lernen und Unterstützte Kommunikation für Schülerinnen und Schüler mit einer körperlichen/geistigen Behinderung: https://www.cluks-forum-bw.de/

COMES:

http://www.comes-care.net/

Communication by gaze interaction (COGAIN)

http://wiki.cogain.org/index.php/Eye_Control_Hints_and_Tips

EHeR versorgt:

Eldercare locator

www.eldercare.gov

Forum der Deutschen Gesellschaft für Muskelkranke e.V. (DGM)

https://www.dgm.org/aktiv-werden/sich-austauschen/foren

GAZE—Functional Gaze Control

https://www.ucl.ac.uk/gaze/gaze-project/downloads

Gesellschaft für Unterstützte Kommunikation e.V.:

http://www.gesellschaft-uk.de

Kinderkanal (2014): „Lennart spricht mit den Augen" über einen Schüler, der mit einer Augensteuerung kommuniziert: http://www.kika.de/schau-in-meine-welt/sendungen/sendung78586.html

National Institute on Aging

www.nia.nih.gov

Pflegewerk

REHADAT: Portal mit Informationen zu Hilfsmitteln

http://www.rehadat-hilfsmittel.de/de/

Smartbox – „10 tips for successfully using eyegaze":

https://www.youtube.com/watch?v=ngYHTNmije8

Talking mats. Symbolbasierte Feedback-Methode

http://www.talkingmats.com/

Vitadock:

https://cloud.vitadock.com/?lang=en

Printed in the United States
By Bookmasters